Green Energy and Technology

For further volumes:
http://www.springer.com/series/8059

Elena V. M. Papadopoulou

Energy Management in Buildings Using Photovoltaics

Springer

Dr. Elena V. M. Papadopoulou
Environmental Technology Laboratory
Department of Mechanical Engineering
University of Western Macedonia
M. Alexandrou 10
50200 Ptolemaida
Greece
e-mail: epapadopoulou@uowm.gr

ISSN 1865-3529   e-ISSN 1865-3537
ISBN 978-1-4471-5896-7   ISBN 978-1-4471-2383-5 (eBook)
DOI 10.1007/978-1-4471-2383-5
Springer London Dordrecht Heidelberg New York

British Library Cataloguing in Publication Data
A catalogue record for this book is available from the British Library

Printed on acid-free paper

Springer is part of Springer Science+Business Media (www.springer.com)

To my "special team" for their patience and support all these days and nights of work

Preface

Energy management in buildings and the optimization of energy efficiency are of continuously growing importance. In today's power scenario, we are facing a major power crunch. Day by day, the gap between demand and supply of electric energy is widening.

Energy management has become an important issue as many utilities around the world find it very difficult to meet energy demands, which has led to load shedding and power quality problems. An efficient energy management in residential, commercial, and industrial sectors can reduce energy requirements and lead to savings in the cost of energy consumed, which also has positive impact on the environment. Energy management is not only important in distribution systems but it also has great significance in generation systems as well. Smart grid management and renewable energy integration are becoming important aspects of efficient energy management.

The aim of this handbook is to turn the attention to these subjects and to promote methods and technologies to increase energy efficiency in buildings reducing the consumption and using photovoltaic energy systems.

The reduction of energy consumption is essential to be considered not only as a manner for the protection of environment via the reduction, for example, of emissions of greenhouse gases but also as a question that concerns the energy management of buildings.

The rapid development of renewable energy in residential areas and the focus on load management open new possibilities of energy management for the final user, in grid-connected multi-source systems with large energy contributions. Such problems meet high uncertainties degrees and computation speed limits and need new tools.

Moreover, this book will develop the use of photovoltaic systems in dwellings, focusing on basic sizes (economically, technically, and administratively) and relevant energy management models. This raises the question of whether such concerns can contribute to energy management in buildings.

The structure of this text allows flexibility in course content and design. It may be used equally well either as a textbook for students or a handbook for engineers who want to introduce themselves to energy management and photovoltaic operation. Coverage of preliminary energy management systems and photovoltaic operations is basic enough for the fundamental photovoltaic course; yet, it is broad and analytical enough to also be used in an advanced course that gives an in-depth treatment of specific topics. Furthermore, this text may be used at either the undergraduate or the graduate level.

To aid the presentation of its subject matter, this text includes the following features:

Figures diagrams and charts to further explain and illustrate the concepts and techniques presented.

Questions at the end of the chapters to provide summaries and reviews of key points. For the most part, these require qualitative answers and may be used to alert the respondent to central ideas which need closer examination.

Design and calculation of an operation of installation of roof-mounted photovoltaic systems.

Glossary with all the terms of photovoltaic technology.

I wish to acknowledge Springer Verlag; with special mentions to Anthony Doyle, Claire Protherough, and Grace Quinn for their editorial support.

Contents

Introduction

Energy management offers the largest and most cost-effective opportunity for both industrialized and developing nations to limit the enormous financial, health, and environmental costs associated with burning fossil fuels. Available, cost-effective investments in energy globally are estimated to be billions of euros per year. However, the actual investment level is far less, representing only a fraction of the existing, financially attractive opportunities for energy savings investments.

Especially for developing countries with rapid economic growth and surging energy consumption, energy- and water-efficient design offers a very cost-effective way to control the exploding costs of building power and water treatment plants, while limiting the expense of future energy imports and the widespread health and environmental damages and costs that result from burning fossil fuels.

Energy PV park (juwi group)

During more than the last decade a continuously increasing interest in renewable energy technologies was noted in European Union. This was a combined effect of: (a) the favorable legal and financial measures that were implemented, (b) the rich potential of renewable energy sources (RES) that exists, and (c) the rising environmental awareness.

Energy efficiency upgrades are far more cost-effective than adding solar panels. Buildings that can demonstrate energy efficiency or will soon undergo efficiency upgrades will be given priority for renewable energy installations.

However, there is misunderstanding on what sustainable/green building is. It is thought that a sustainable building should be an expensive high-tech assembling, and some pilot projects are of this kind, which results in more investment, more energy consumption, and more environmental impact.

Green building (also known as *green construction* or *sustainable building*) refers to a structure and using process that is environmentally responsible and resource-efficient throughout a building's life cycle: from sitting to design, construction, operation, maintenance, renovation, and demolition. This practice expands and complements the classical building design concerns of economy, utility, durability, and comfort.

Although new technologies are constantly being developed to complement current practices in creating greener structures, the common objective is that green buildings are designed to reduce the overall impact of the built environment on human health and the natural environment by:

- efficiently using energy, water, and other resources
- protecting occupant health and improving employee productivity
- reducing waste, pollution, and environmental degradation

A similar concept is natural building, which is usually on a smaller scale and tends to focus on the use of natural materials that are available locally. Other related topics include sustainable design and green architecture. Sustainability may be defined as meeting the needs of present generations without compromising the ability of future generations to meet their needs. Green building does not specifically address the issue of the retrofitting existing homes.

However, modern sustainability initiatives call for an integrated and synergistic design to both new construction and in the retrofitting of an existing structure. Also known as sustainable design, this approach integrates the building life cycle with each green practice employed with a design purpose to create a synergy among the practices used.

Green building brings together a vast array of practices and techniques to reduce and ultimately eliminate the impacts of buildings on the environment and human health. It often emphasizes taking advantage of renewable resources, e.g., using sunlight through passive solar, active solar, and photovoltaic techniques and using plants and trees through green roofs, rain gardens, and for reduction of rainwater run-off. Many other techniques, such as using packed gravel or permeable concrete instead of conventional concrete or asphalt to enhance replenishment of ground water are used as well.

On the aesthetic side of green architecture sustainable design is the philosophy of designing a building that is in harmony with the natural features and resources surrounding the site. There are several key steps in designing sustainable buildings: specify 'green' building materials from local sources, reduce loads, optimize systems, and generate on-site renewable energy.

Green buildings often include measures to reduce energy consumption—both the embodied energy required to extract, process, transport and install building materials and operating energy to provide services such as heating and power for equipment.

As high-performance buildings use less operating energy, embodied energy has assumed much greater importance—and may make up as much as 30% of the overall life cycle energy consumption. To reduce operating energy use, high-efficiency windows and insulation in walls, ceilings, and floors increase the efficiency of the building envelope, (the barrier between conditioned and unconditioned space). Another strategy, passive solar building design, is often implemented in low-energy homes. Designers orient windows and walls and place awnings, porches, and trees to shade windows and roofs during the summer while maximizing solar gain in the winter. In addition, effective window placement (day lighting) can provide more natural light and lessen the need for electric lighting during the day. Solar water heating further reduces energy costs.

On-site generation of renewable energy through solar power, wind power, hydro power, or biomass can significantly reduce the environmental impact of the building. Power generation is generally the most expensive feature to add to a building.

Chapter 1
Energy Management

The effective management of energy costs has gained an increasing importance because of the difficulties that they faced by the recession which followed the international energy and financial crisis.

However, there are additional reasons that require buildings to be more aware of energy saving. The global reserves of energy raw materials are getting dry. At the same time, international efforts to reduce the wasted energy which is responsible for adverse climatic changes become more intense.

Commercial and residential buildings are the largest consumers of electricity and contribute significantly to greenhouse gas emissions. As a result, building energy management schemes are being deployed to reduce/manage building energy use; reduce electricity bills while increasing occupant comfort and productivity; and improve environmental stewardship without adversely affecting standards of living. The attainment of these energy management goals requires insight into appliance usage patterns and individual appliance energy use, combined with intelligent appliance operation and control (Fig. 1.1).

The International Energy Association also predicts that electricity usage for residential appliances would grow by 25% by 2020.

Building owners are fully aware of the need for effective protection of the environment. They will increasingly prefer using buildings that provide services and products more environmentally friendly and contribute to the saving of valuable energy sources.

Energy is a mainstay of modern industrial society. The energy sources on which the European energy infrastructure is based are mostly fossil fuels. The energy use can be divided into three end use segments: transportation, buildings and industrial ones. Each of these sectors consumes about one-third of the total energy use. The energy consumption in the building sector represents almost 40% of the total energy consumption in Europe whilst contributing significantly to the greenhouse gas emissions.

E. V. M. Papadopoulou, *Energy Management in Buildings Using Photovoltaics*, Green Energy and Technology, DOI: 10.1007/978-1-4471-2383-5_1, © Springer-Verlag London Limited 2012

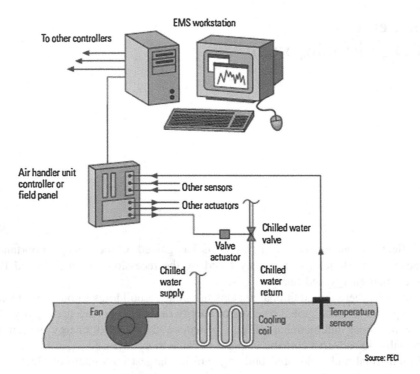

Fig. 1.1 EMS workstation design

On the other hand, there is increasing international concern with climate change, and the targets agreed by the European Union under the Kyoto Protocol to reduce emissions of greenhouse gases.

The complexity of systems deployed on modern buildings, necessitates the use of optimal control. During the last years, there is a rapid convergence of the technologies of Informatics, Microelectronics and Control Systems leading to novel approaches and solutions for energy and building automation related problems.

As the 'intelligent building' is passing nowadays its phase of maturity, a great number of manufacturers offer integrated solutions. The fault detection and diagnosis (FDD) technology provides the capability to deal with complex problems that are related with the uninterrupted operation of various systems even in a fault regime.

Energy Management of Buildings is the judicious and effective use of energy. The most important are to save energy, to minimize energy costs, to waste without affecting living quality and finally to minimize environmental effects. Energy management systems control energy-consuming building equipment to make it operate more efficiently and effectively. On average, energy management systems save about 10% of overall annual building energy consumption.

Fig. 1.2 Energy
management system eclipse

An *Energy Management System* (EMS) controls how energy is consumed in a building and how building equipment operates. Systems that meet this definition may vary widely in sophistication, from simple time clocks that switch equipment on and off to advanced computerized controls that provide centralized equipment management (Fig. 1.2).

EMS is a system of computer-aided tools used by operators of electric utility grids to monitor, control, and optimize the performance of the generation and/or transmission system. The monitor and control functions are known as Supervisory Control and Data Acquisition (SCADA); the optimization packages are often referred to as "advanced applications".

The computer technology is also referred to as SCADA/EMS or EMS/SCADA. In these respects, the terminology EMS then excludes the monitoring and control functions, but more specifically refers to the collective suite of power network applications and to the generation control and scheduling applications.

Manufacturers of EMS also commonly supply a corresponding dispatcher training simulator (DTS). This related technology makes use of components of SCADA and EMS as a training tool for control center operators. It is also possible to acquire an independent DTS from a non-EMS source.

The evolution of automatic building control began in the 1880s. The first innovation was a bimetal-based thermostat with a hand-wound spring-powered motor which controlled space temperature by adjusting a draft damper on a coal-fired furnace or boiler. In 1890, the first pneumatic powered control became available.

Today automated energy control has become standard practice. Virtually all nonresidential buildings have automatic controllers with a computer as the central processor. These systems are called EMS, Energy Management Control Systems (EMCS), or Building Automation Systems (BAS). Building owners and facility managers must regularly address the issue of computerized energy management—assessing existing systems, specifying and commissioning new systems, evaluating service contract options, or optimizing EMS operations.

The term EMS can also refer to a computer system which is designed specifically for the automated control and monitoring of those electromechanical facilities in a building which yield significant energy consumption such as heating, ventilation and lighting installations. The scope may span from of a single building to a group of buildings such as university campuses, office buildings, retail stores networks or factories. Most of these energy management systems also provide facilities for the reading of electricity, gas and water meters. The data obtained from these can then be used to perform self-diagnostic and optimization routines on a frequent basis and to produce trend analysis and annual consumption forecasts.

EMSs are capable of saving, on average, about 10% of overall annual building energy consumption. In general, energy management systems save and manage energy by controlling equipment so that:

- equipment is running only when necessary;
- equipment is operating at the minimum capacity required; and
- peak electric demand is minimized.

EMSs also may be used to save energy by monitoring equipment operational data, which may then be used for diagnostics and troubleshooting.

A *Building Management System* (BMS) is a computer-based control system installed in buildings that controls and monitors the building's mechanical and electrical equipment such as ventilation, lighting, power systems, fire systems, and security systems. A BMS consists of software and hardware; the software program, usually configured in a hierarchical manner, can be proprietary, using such protocols as C-bus, Profibus, and so on. The Building Energy Management Systems (BEMS) market, a growing segment of the larger building efficiency industry, is gaining momentum as an alternative, cheaper means for end-users to implement energy efficiency applications in commercial buildings. Conceived out of the market demand for a lighter, less-expensive and strictly energy-related automation and management systems for commercial buildings, the BEMS market includes both specialized and broad-based solutions. Such solutions range from reactive energy efficiency optimization software to predictive supply and demand side energy management architectures.

The BEMS market will continue to grow at a strong pace over the next few years, as players in adjacent markets continue to invest in the space. These players will include IT vendors, BMS vendors, curtailment service providers (CSPs), and other energy efficiency companies. As a result of market convergence, BEMS offerings will become more sophisticated, providing energy savings to the end-user that in many cases will be reinvested in additional energy efficiency applications.

Additionally, the return on investment for BEMS will become even more attractive as costs are driven down—as BEMS vendors enter into more buildings, implementation templates will begin to take shape. Looking ahead to 2016, many end-users with portfolio management needs will choose BEMS rather than updating a BMS, as it is a less-expensive option, and BEMS are technology agnostic (Fig. 1.3).

A BMS is most common in a large building. Its core function is to manage the environment within the building and may control temperature, carbon dioxide levels and humidity within a building. As a core function in most BMS systems, it controls heating and cooling, manages the systems that distribute this air throughout the building (for example by operating fans or opening/closing

Fig. 1.3 BEMS control

dampers), and then locally controls the mixture of heating and cooling to achieve the desired room temperature. A secondary function sometimes is to monitor the level of human-generated CO_2, mixing in outside air with waste air to increase the amount of oxygen while also minimizing heat/cooling losses.

Systems linked to a BMS typically represent 40% of a building's energy usage; if lighting is included, this number approaches 70%. BMS systems are a critical component to managing energy demand. Improperly configured BMS systems are believed to account for 20% of building energy usage, or approximately 8% of total energy usage in the United States.

As well as controlling the building's internal environment, BMS systems are sometimes linked to access control (turnstiles and access doors controlling who is allowed access and egress to the building) or other security systems such as closed-circuit television (CCTV) and motion detectors. Fire alarm systems and elevators are also sometimes linked to a BMS, for example, if a fire is detected then the system could shut off dampers in the ventilation system to stop smoke spreading and send all the elevators to the ground floor and park them to prevent people from using them in the event of a fire.

To create a central computer controlled method which has three basic functions:

- Controling
- Monitoring
- Optimizing

the building's facilities, mechanical and electrical equipment for comfort, safety and efficiency.

The most common benefits using BMS are:

- Good control of internal comfort conditions
- Possibility of individual room control
- Increased staff productivity
- Effective monitoring and targeting of energy consumption

- Improved plant reliability and life
- Effective response to HVAC-related complaints
- Save time and money during the maintenance
- Higher rental value
- Flexibility on change of building use
- Individual tenant billing for services facilities manager
- Central or remote control and monitoring of building
- Increased level of comfort and time saving

The field of BMS encompasses an enormous variety of technologies, across commercial, industrial, institutional and domestic buildings, including energy management systems and building controls. The function of BMS is central to 'Intelligent Buildings' concepts; its purpose is to control, monitor and optimize building services, e.g., lighting; heating; security, CCTV and alarm systems; access control; audio-visual and entertainment systems; ventilation, filtration and climate control, etc.; even time and attendance control and reporting (notably staff movement and availability).

The potential within these concepts and the surrounding technology is vast, and our lives are changing from the effects of Intelligent Buildings developments on our living and working environments.

The origins of Intelligent Buildings and BMS have roots in the industrial sector in the 1970s, from the systems and controls used to automate production processes and to optimize plant performances. The concepts and applications were then adapted, developed and modularized during the 1980s, enabling transferability of the technology and systems to the residential and commercial sectors.

The essence of BMS and Intelligent Buildings is in the control technologies, which allow integration, automation, and optimization of all the services and equipment that provide services and manages the environment of the building concerned.

Especially in commercial buildings there problems that contribute to higher than expected energy use in is the lack of actionable data and analysis tools to link control strategies, operations and energy use in the built environment. Conventional building operations are subject to several problems. Monitoring and diagnostics systems rely on a variety of measured data sources to gain insights to the actual building performance, with limited understanding of the information uncertainty and lacking a simple and actionable operator interface. Such limitations result in the inability to diagnose and have corrective actions when the building or its systems are not behaving as expected. Use of fixed schedules and equipment set points (based on equipment performance optimization) limit the ability to achieve overall energy use reduction (Fig. 1.4).

Programmable Logic Controllers (PLC's) formed the original basis of the control technologies.

Later developments, in commercial and residential applications, were based on 'distributed-intelligence microprocessors'.

Fig. 1.4 PLC board

The use of these technologies allows the optimization of various site and building services, often yielding significant cost reductions and large energy savings. There are numerous methods by which building services within buildings can be controlled, falling broadly into two method types:

- *Time-based*—providing heating or lighting services, etc., only when required, and
- *Optimizer parameter-based*—often utilizing a representative aspect of the service, such as temperature for space heating or illuminance for lighting.

Time-based controls can be used to turn on and off the heating system (and/or water heating) at pre-selected periods (of the day, of the week, etc.).

Optimiser parameters, whatever the conditions, the controls make sure the building reaches the desired temperature when occupancy starts.

Temperature control protection against freezing or frost protection generally involves running heating system pumps and boilers when external temperature reaches a set level (0°C).

Compensated systems will control flow temperature in the heating circuit relative to external temperature. This will give a rise in the circuit flow temperature when outside temperature drops.

Thermostatic radiator valves sense space temperature in a room and throttle the flow accordingly through the radiator or convector to which they are fitted.

Proportional control involves switching equipment on and off automatically to regulate output.

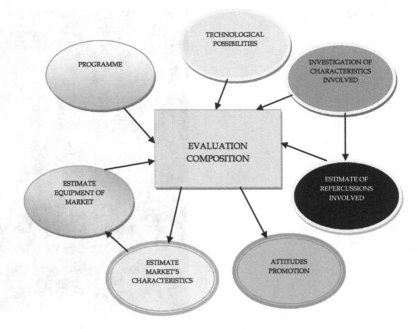

Fig. 1.5 PLC board

Other methods can include thermostats, occupancy sensing passive infra-red sensors (PIR's), and manual user control.

Predictive control of chilled water: evaluate the feasibility of and to demonstrate the energy savings potential of model predictive control (MPC) for set-point optimization and scheduling of a district cooling system with thermal storage serving

Energy performance visualization system: demonstrate real-time energy performance visualization capability for the building (COB).

Occupancy-based energy management system: real-time knowledge of actual building occupancy and contrast with traditional CO_2 sensor-based demand controlled ventilation strategies.

Lighting control methods: time-based control and optimizer parameter-based where a level of illuminance or particular use of lighting is required. Especially:

- Zones: lights are switched on corresponding to the use and layout of the lit areas, in order to avoid lighting a large area if only a small part of it needs light.
- Time control: to switch on and off automatically in each zone to a preset schedule for light use.
- PIR occupancy sensing: in areas which are occupied intermittently, occupancy sensors can be used to indicate whether or not anybody is present and switch the light on or off accordingly.

- Light level monitoring: this consists of switching or dimming artificial lighting to maintain a light level measured by a photocell (Fig. 1.5).

Questions

1. Define the Energy Management?
2. What is the objective of Energy Management?
3. What are the principles of Energy Management?
4. What is the definition of BEMS?
5. What is the difference between BEMS and legacy BMS?

Chapter 2
Energy Efficiency and Energy Saving

Until recent years, energy efficiency has been a relatively low priority and low perceived opportunity to building owners and investors. However, with the dramatic increase and awareness of energy use concerns, and the advances in cost-effective technologies, energy efficiency is fast becoming part of real estate management, facilities management, and operations strategy. The concepts are also now making significant inroads into the domestic residential house building sectors.

For lighting, energy savings can be up to 75% of the original circuit load, which represents 5% of the total energy consumption of the residential and commercial sectors.

Energy savings potential from water heating, cooling, or hot water production, can be up to 10%, which represents up to 7% of the total energy consumption of the domestic residential and commercial sectors.

Model Predictive Control (MPC) offers energy saving potential in large buildings and systems for which response to external disturbances or control inputs is slow, i.e. on the order of hours. MPC effectively provides a means to optimize systems dynamically to take advantage of building utilization, weather patterns, and utility rate structures.

Dynamic system models could be used to identify critical control variables (from an energy performance standpoint), guide facility operation, and produce a predictive controller that reduces energy use. HVAC systems and building hydronic systems are becoming prevalent and robust; multivariable control methods are lacking. With the visualization prototype providing comparison of performance metrics to previous years, building models, and benchmarks, the facility manager can assess the energy and cost savings of a particular action with relative certainty; a traditional BMS may not store data for the length of time necessary to provide such insight, nor does it provide relevant benchmarks or models to show how the building is expected to perform (Table 2.1 and Fig. 2.1).

E. V. M. Papadopoulou, *Energy Management in Buildings Using Photovoltaics*,   11
Green Energy and Technology, DOI: 10.1007/978-1-4471-2383-5_2,
© Springer-Verlag London Limited 2012

Table 2.1 Data requirement and calculation procedures of Building Performance Metrics

Metric	Unit	Data	Calculation
Total electricity consumption	kWh/year/ gsf	• kW data every time step,for a whole year •Total gross sq.ft. (incl.the wall)	Sum (kW data at every time step
Electricity demand	kW/gsf	Same as above	Max (kW data at every time step)
Total gas consumption	therms/ gsp-year	•Thermes/h hot water data at every time step, for a whole year •Total gross sq.ft. (incl.the wall) •Central boiler plant efficiency	Sum (therms/h gas consumption data at every time step)
Gas demand	therms/ h-gsf-year	Same as above	Max (therms/h gas consumption at every time step)

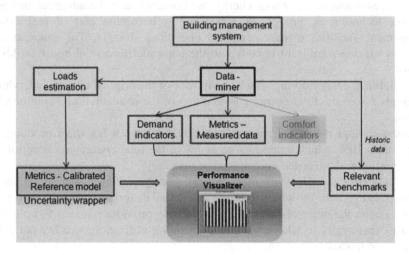

Fig. 2.1 BMS typical schedule

Savings are anticipated to be higher in buildings where an extensive CO_2 sensor network is present.

For designing the energy plant for buildings, we have to know the energy requirements and the energy demand. Energy requirement is the amount of energy needed in order to meet requirements for indoor temperature, indoor humidity, illuminance etc.

Energy demand is the amount of energy the systems present must use in order to meet energy requirements. The type of energy demand required for benefits delivery, distribution, and energy generation is known as energy demand for generation or final energy demand.

Fig. 2.2 Energy demand in
Europe

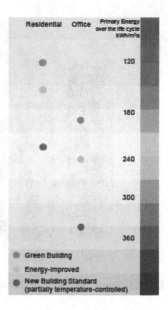

Energy demand defines the amount of energy that is used from energy supply companies (oil, gas, wood, electricity etc.). Related to final energy demand, we also have primary energy demand, the so-called prechain for primary energy sources used: exploration, extraction, harvesting, transport and conversion— basically, the entire path to the generating system inside the building. Each primary energy source is allocated a so-called primary energy factor that tracks all energy-relevant demand from initial harvesting to the building limits.

For new buildings and redevelopment projects, primary energy demand for any systems used for conditioning a room must, depending on utilization, stay below certain critical values. Implementation of these Europe-wide requirements was a big step in the right direction, for once, but it does not suffice.

Energy-based evaluation of energy-efficient and sustainable buildings must take into account all energy flows including, for instance, energy demand for the manufacture, renewal and maintenance of building materials as well as electricity demand for furnishings supplied by the occupant (Fig. 2.2).

2.1 Greenhouse Gas Benefits

A greenhouse gas (sometimes abbreviated GHG) is a gas in an atmosphere that absorbs and emits radiation within the thermal infrared range. This process is the fundamental cause of the greenhouse effect.

Europe has made remarkable progress in reducing the energy and carbon intensity of its building stock and operations. Energy use in buildings since 1972

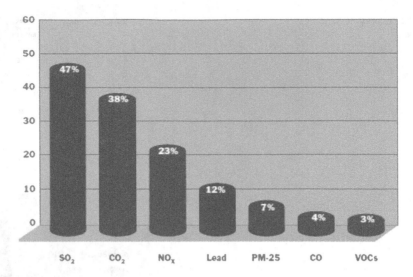

Fig. 2.3 Energy emissions from buildings

has increased at less than half rate of growth of Europe's gross domestic product, despite the growth in home size and building energy services such as air-conditioning and consumer and office electronic equipment. Although great strides have been made, abundant untapped opportunities still exist for further reductions in energy use and emissions. Many of these—especially energy-efficient building designs and equipment—would require only modest levels of investment and would provide quick pay-back to consumers through reduced energy bills. By exploiting these opportunities, Europe could have a more competitive economy, cleaner air, lower GHG emissions, and greater energy security.

Greenhouse gas emission reductions depend on and correlate to reductions in energy use.

GHG emissions from the building sector in EU have been increasing at almost 2% per year since 1990, and CO_2 emissions from residential and commercial buildings are expected to continue to increase at a rate of 1.4% annually through 2025. These emissions come principally from the generation and transmission of electricity used in buildings, which account for 71% of the total. Due to the increase in products that run on electricity, emissions from electricity are expected to grow more rapidly than emissions from other fuels used in buildings. In contrast, direct combustion of natural gas (e.g., in furnaces and water heaters) accounts for about 20% of energy-related emissions in buildings, and fuel-oil heating in the Northeast and Midwest accounts for the majority of the remaining energy-related emissions. Based on energy usage, opportunities to reduce GHG emissions appear to be most significant for space heating, air-conditioning, lighting, and water heating (Fig. 2.3).

The avoided damages of climate change are estimated for a range of emission reduction policies from a range of business as usual scenarios. In the emission

Fig. 2.4 Typical daily power consumption

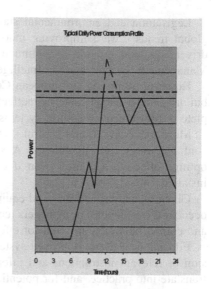

abatement scenarios, concentrations of greenhouse gases overshoot before falling to a stable level. The peak concentrations are used to characterize the stabilization scenario.

Similarly, the peak impacts are used to evaluate the scenarios. This is in line with avoiding "dangerous interference with the climate system". Results are shown for both cost-effective and "realistic" emission reduction policies.

Avoided climate change impacts increase with emission abatement, but the additionally avoided impacts fall as abatement gets more stringent. The most serious climate change impacts can be avoided with only modest emission reduction. Very stringent emission reduction may even increase climate change impacts because of the removal of the sulfur veil and because emission abatement costs may slow economic growth and increase vulnerability.

Most electricity is generated by fossil fuel-fired power stations. These release carbon dioxide (a greenhouse gas) into the atmosphere, as a by-product of electricity generation. Home energy use is responsible for 30% of USA's carbon dioxide emissions and for 40% of EU's, which contribute to climate change.

By following best practice standards, newly built and refurbished housing will be more energy efficient: this will reduce emissions, cut energy consumption, save money, and safeguard the environment.

New properties are built to high thermal performance standards and so require less heating. As a result, lighting and appliances can account for up to 75% of total fuel costs, 33% of CO_2 emissions, and about 20% of total energy use (Fig. 2.4).

Intelligent Buildings and Building Management Systems technologies contribute directly to the reduction in energy use in commercial, industrial, institutional, and domestic residential sectors.

In short, Intelligent Buildings and suitably applied Building Management Systems are good for the environment.

Legislation and environmental standards; health and safety regulations; and global trends toward improving indoor air quality standards are all significant drivers of—and provide a continuous endorsement of the need for—Building Management Systems and the Intelligent Buildings technologies.

However according to European Commission as many as 90% of all existing buildings have inapplicable or ineffective controls, many of which require complete refurbishment of control systems.

Moreover conventional control systems stop when short of automated Intelligent Buildings full capabilities. A significant human element is required for optimal effective operation even if control systems are correctly specified and installed.

Given typical installations and equipment there is often a difficulty for building occupants (residential) or managers (commercial) to operate them correctly. Usage and correct operation are vital for effective results.

Education of users; improved systems-design user-friendliness, and the provision of relevant instructions and information are all critical to enable theory to translate into practice, and for potential effectiveness and savings to be realized.

Energy-effective systems balance a building's electric light, daylight, and mechanical systems for maximum benefit.

Enhanced lighting design is more than an electrical layout. It must consider the needs and schedules of occupants, seasonal and climatic daylight changes, and its impact on the building's mechanical systems.

2.2 Lighting and Mechanical Systems

Controlled lighting and mechanical systems can save much energy and minimize building's energy consumption.

Lighting systems consist of ballasts and luminaires or lighting fixtures. Ballasts provide the start-up voltages required for lamp ignition, and regulate current flow through the bulb. Newer ballasts enable fluorescent dimming using analog or digital methods, enabling granular control of lighting output. It has been discovered that the human eye is insensitive to dimming of lights by as much as 20%, as long as the dimming is performed at a slow enough rate (Akashi and Neches 2004), thereby permitting significant savings in energy use.

Energy-efficient lighting will reduce electricity consumption.

Basic lighting control modes include on/off control, scheduling, occupancy detection, and dimming. More advanced schemes include daylight harvesting, task tuning, and demand response. Daylight harvesting involves measurement of how much ambient light is present, and harnessing ambient light to reduce the amount of artificial lighting required to keep the amount of light at a preset level. Task tuning involves adjusting the light output in accordance with the function or tasks which will be performed in the lighted area. Demand response is the dimming of lighting output in response to signals from the utility. Intelligent lighting control systems combine digital control with computation and communications capabilities.

Adding daylight to a building is one way to achieve an energy-effective design. Natural daylight 'harvesting' can make people happier, healthier, and more productive. And with the reduced need for electric light, a great deal of money can be saved on energy. Nearly every commercial building is a potential energy saving project, where the electric lighting systems can be designed to be dimmed with the availability of daylight. Up to 75% of lighting energy consumption can be saved. In addition, by reducing electric lighting and minimizing solar heat gain, controlled lighting can also reduce a building's air-conditioning load.

Energy consumption can be reduced by:

- Using energy-efficient lamps and luminaires (light fittings)
- Directing light where needed (task-directional lighting)
- Using lighting only when required
- Making the most of daylight.

Energy-efficient lighting can:

- Reduce energy costs
- Reduce CO_2 emissions
- Reduce maintenance costs for communal areas through longer life expectancy
- Help to achieve Housing Corporation Scheme Development Standards
- Aid compliance with building regulations.

The best time to install low energy lighting is during construction. In existing housing, luminaires can be replaced at any time. A key opportunity, though, is while rewiring—especially where existing pendant fittings are being replaced. In unoccupied housing (e.g. during a tenancy gap, or redevelopment) low energy lighting should be installed as part of the works.

Incandescent lamps are very inefficient (90% of the energy they use is given out as heat). Because of this, though, they help to keep a building warm during the heating season. Changing to energy-efficient lighting may therefore mean more energy is needed for heating (this is known as the heat replacement effect). The additional energy requirement will partially offset the cost and CO_2 savings of energy-efficient lighting (Table 2.2).

2.3 HVAC System

The HVAC system and controls, including the distribution system of air into the workspaces, are the mechanical parts of buildings that affect thermal comfort (Fig. 2.5).

Table 2.2 CFL versus WGLS

Comparing compact fluorescent lamps (CFL) and general lighting service (GLS) lamps		
	100 W GLS	20 W CFL
Cost (euros)	0.57	4.20
Lamp life (hours)	1,000	12,000
Total lamp cost (over life of 1 CFL) (euros)	6.61	4.20
Total electricity cost (over life of 1 CFL) (euros)	107.57	21.51
Total costs (euros)	114.37	25.71
Savings through the use of CFL instead of 100 W GLS (euros)		88.663

Fig. 2.5 Typical conventional heating/cooling system

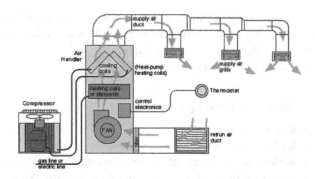

These systems must work together to provide building comfort. While not usually a part of the aesthetics of a building, they are critical to its operations and occupant satisfaction.

The number one office complaint is that the workplace is too hot. Number two is that it is too cold.

Many people cope by adding fans, space heaters, covering up vents, complaining, conducting 'thermostat wars' with their co-workers, or simply leaving the office. Occupants can be driven to distraction trying to adjust the comfort in their space. Improper temperature, humidity, ventilation, and indoor air quality can also have significant impacts on productivity and health. When we are thermally comfortable we work better, shop longer, relax, breathe easier, and focus our attention better.

In order to provide a comfortable and healthy indoor environment the building mechanical system must:

- Provide an acceptable level of temperature and humidity and safe guard against odors and indoor air pollutants
- Create a sense of habitability through air movement, ventilation, and slight temperature variation
- Allow the occupant to control and modify conditions to suit individual preferences.

2.4 BMS for Residential Applications

Many factors influence how well someone feels at home. Among these are a comfortable temperature, good air quality, and the most natural and glare-free light possible. An energy-optimized building should offer a year-round comfortable living conditions and low energy demand. For example, if a modularly constructed façade system replaces the outside wall of a building it will guarantee at the same time the best possible supply of daylight and fresh air. In effect, this synergy façade combines the function of the building's walls with the tasks of household technology.

With the widespread adoption of digital technologies there will be a profound change in how we communicate with others.

Until recently, the majority of homes were wired with little more than the main electrical circuits, a few phone lines, and a few TV cables. Times have changed. Electrical and security system contractors routinely install low voltage communication network cables for a wide range of intelligent home or 'smart home' systems.

Services and equipment that utilize these networks include: security; home theater and entertainment; telephones, door-phones and intercoms; PC and internet networks; surveillance cameras; driveway vehicle sensors; communicating thermostats; motorized window blinds and curtains; entry systems; and irrigation systems.

'Smart home' is an alternative term for an intelligent residential building, or an intelligent home.

Integrating the home systems allows them to communicate with one another through the control system, thereby enabling single button and voice control of the various home systems simultaneously, in preprogrammed scenarios or operating modes.

The development of smart home systems focus on how the house and its related technologies, products, and services should evolve to best meet the opportunities and challenges of the future. The possibilities and permutations are endless.

In a "smart building," the systems, structure, and function of the building are optimized to provide a productive and cost-effective environment that adapts to the current and future needs of building occupants. Although smart buildings are diverse in nature and employ many different techniques to enhance productivity and drive down cost, most smart buildings share a set of characteristics that set them apart from other buildings.

Building energy management is the process of monitoring and controlling the operating systems within a building. Though specific components may differ, these operating systems may include heating and air-conditioning, ventilation, lighting, power, security, and alarm systems. While building energy management techniques can be applied to a variety of building types, they are generally the most cost-effective when used in large commercial and industrial buildings. The International Energy Conservation Code (IECC) provides codes and standards that are often used in developing building energy monitoring standards at the local level.

Most building energy management systems are operated using specially designed software programs. These programs are operated using a traditional computer, and are capable of providing feedback on system operations and energy consumption. Most types of building energy management software also allow operators to make changes to building automation systems, though some may require changes to be made manually. These energy control systems are usually operated by building management or maintenance personnel, who must be trained to interpret the building performance data generated by the software. Many building management have both hands-on and technical training in energy conservation and building operations.

The trending and monitoring capabilities of EMS are powerful tools for improving heating, ventilation, air-conditioning (HVAC) and lighting and for reducing energy use, but most facility managers and system operators simply do not have the time to fully investigate these resources. Those responsible for EMS upgrade or purchase are not always able to study their facility's exact energy management needs themselves; they may rely on vendors to provide specifications—and they may not receive the optimal system for their building. Furthermore, the commissioning process, which can be critical to the success of an EMS, is relatively unknown to most facility staff.

Questions

1. What is the total revenue opportunity for BEMS?
2. What are the total potential BEMS revenues for the three major verticals: commercial, university, and health care?
3. How energy-efficient lighting reduces energy consumption?
4. What offers a model predictive control (MPC)?
5. Where could the dynamic system models be used?

Chapter 3
Zero-Energy Buildings

A *zero-energy building*, also known as a zero net energy (ZNE) building, is a popular term to describe a buildings use with zero net energy consumption and zero carbon emissions annually.

Zero-energy buildings can be used autonomously from the energy grid supply—energy can be harvested on-site usually in combination with energy-producing technologies such as solar and wind while reducing the overall use of energy with extremely efficient HVAC and lighting technologies. The zero-energy design principle is becoming more practical in adopting due to the increasing costs of traditional fossil fuels and their negative impact on the planet's climate and ecological balance.

The zero net energy consumption principle is gaining considerable interest as renewable energy harvesting as a means to cut greenhouse gas emissions. Traditional building use consumes 40% of the total fossil energy in the US and European Union. In developing countries many people have to live in zero-energy buildings out of necessity. Many people live in huts, yurts, tents, and caves exposed to temperature extremes and without access to electricity. These conditions and the limited size of living quarters would be considered uncomfortable in the developed countries.

The development of modern zero-energy buildings became possible not only through the progress made in new construction technologies and techniques, but it has also been significantly improved by academic research on traditional and experimental buildings, which collected precise energy performance data. Today's advanced computer models can show the efficacy of engineering design decisions.

Energy use can be measured in different ways (relating to cost, energy, or carbon emissions) and, irrespective of the definition used, different views are taken on the relative importance of energy harvest and energy conservation to achieve a net energy balance. Although zero-energy buildings remain uncommon in developed countries, they are gaining importance and popularity. The zero net energy

E. V. M. Papadopoulou, *Energy Management in Buildings Using Photovoltaics*,
Green Energy and Technology, DOI: 10.1007/978-1-4471-2383-5_3,
© Springer-Verlag London Limited 2012

Fig. 3.1 Zero-energy
building approach

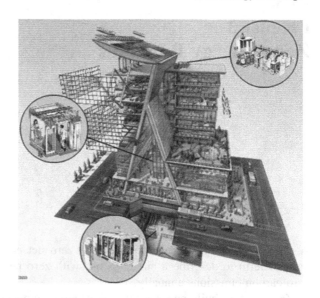

approach has potential to reduce carbon emissions, and reduce dependence on
fossil fuels.

A building approaching zero net energy use may be called a "near-zero energy
building" or "ultra-low energy house". Buildings that produce a surplus amount of
energy during a portion of the year may be known as "energy-plus buildings"
(Fig. 3.1).

If the building is located in an area that requires heating or cooling throughout
parts of the year, it is easier to achieve zero net energy consumption when the
available living space is kept small.

Advanced HVAC and lighting controls are the "brains" behind the intelligent
operation of zero-energy buildings. Controls that improve the intelligent response
of building systems include the following characteristics:

- Direct digital controls (DDCs) with electronic ancillary devices for both central
  equipment control and zone-level management
- Integration into one central BAS
- Interoperability achieved through an open protocol standard

The benefits of a ZEB usually are as follows:

- Isolation for building owners from future energy price increases
- Increased comfort due to more uniform interior temperatures
- Reduced requirement for energy austerity
- Reduced total cost of ownership due to improved energy efficiency
- Reduced total net monthly cost of living
- Improved reliability—photovoltaic systems have 25-year warranties—seldom
  fail during weather problems

- Extra cost is minimized for new construction compared to an afterthought retrofit
- Higher resale value as potential owners demand more ZEBs than available supply
- The value of a ZEB relative to similar conventional building should increase every time energy costs increase
- Future legislative restrictions, and carbon emission taxes/penalties may force expensive retrofits to inefficient buildings.

The disadvantages of a ZEB will be as:

- Initial costs can be higher—effort required to understand, apply, and qualify for ZEB subsidies
- Very few designers or builders have the necessary skills or experience to build ZEBs
- Possible declines in future utility company renewable energy costs may lessen the value of capital invested in energy efficiency
- New photovoltaic solar cells equipment technology price has been falling at roughly 17% per year—it will lessen the value of capital invested in a solar electric generating system—current subsidies will be phased out as photovoltaic mass production lowers future price
- Challenge to recover higher initial costs on resale of building—appraisers are uninformed—their models do not consider energy
- Climate-specific design may limit future ability to respond to rising-or-falling ambient temperatures (global warming)
- While the individual house may use an average of net zero energy over a year, it may demand energy at the time when peak demand for the grid occurs. In such a case, the capacity of the grid must still provide electricity to all loads. Therefore, a ZEB may not reduce the required power plant capacity
- Without an optimized thermal envelope the embodied energy, heating and cooling energy and resource usage is higher than needed. ZEB by definition does not mandate a minimum heating and cooling performance level thus allowing oversized renewable energy systems to fill the energy gap
- Solar energy capture using the house envelope only works in locations unobstructed from the South. The solar energy capture cannot be optimized in south facing shade or wooded surroundings.

The most cost-effective steps toward a reduction in a building's energy consumption usually occurs during the design process. To achieve efficient energy use, zero-energy design departs significantly from conventional construction practice. For a successful zero energy building design we have to combine time-tested passive solar, or natural conditioning, principles that work with the on-site assets. Sunlight and solar heat, prevailing breezes, and the cool of the earth below a building, can provide day lighting and stable indoor temperatures with minimum mechanical means.

Fig. 3.2 Zero-energy
dwellings

ZEBs are normally optimized to use passive solar heat gain and shading, combined with thermal mass to stabilize diurnal temperature variations throughout the day, and in most climates are super insulated.

All the technologies needed to create zero-energy buildings are available off-the-shelf today. A zero-energy building is capable of providing tenants with all of these desired building features (Fig. 3.2).

Advanced HVAC control systems enable greater tenant control of temperature, ensure more comfortable temperatures throughout the building, monitor indoor air quality, and give facilities managers the information and control they need to quickly respond to tenant complaints. Smart buildings are also designed to allow easy reconfiguration of suite layout (Fig. 3.3).

Sophisticated 3D computer simulation tools are available to model how a building will perform with a range of design variables such as building orientation (relative to the daily and seasonal position of the sun), window and door type and placement, overhang depth, insulation type and values of the building elements, air tightness (weatherization), the efficiency of heating, cooling, lighting, and other equipment, as well as local climate. These simulations help the designers predict how the building will perform before it is built, and enable them to model the economic and financial implications on building cost benefit analysis, or even more appropriate—life cycle assessment.

Zero-energy buildings are built with significant energy-saving features.

The heating and cooling loads are lowered by using high-efficiency equipment, added insulation, high-efficiency windows, natural ventilation, and other techniques. These features vary depending on climate zones in which the construction occurs. Water heating loads can be lowered by using water conservation fixtures, heat recovery units on waste water, and by using solar water heating, and high-efficiency water heating equipment.

In addition, day lighting with skylites or solartubes can provide 100% of daytime illumination within the home. Nighttime illumination is typically done with fluorescent and LED or high powered LED lighting that use one-third or less power than incandescent lights, without adding unwanted heat. And miscellaneous electric loads can be lessened by choosing efficient appliances and minimizing phantom loads or standby power.

Fig. 3.3 Zero energy house

Other techniques to reach net zero (dependent on climate) are Earth sheltered building principles, super insulation walls using straw-bale construction, Vitruvian built pre-fabricated building panels and roof elements plus exterior landscaping for seasonal shading.

Zero-energy buildings share a set of characteristics that set them apart from other buildings. These characteristics include:

- Advanced HVAC and lighting controls
- Smart metering capabilities, allowing central access to real-time utility data
- A structured cabling infrastructure with high bandwidth and connectivity
- Adaptability to changing technology and tenant needs

Zero-energy buildings are often designed to make dual use of energy including white goods; for example, using refrigerator exhaust to heat domestic water, ventilation air and shower drain heat exchangers, office machines and computer servers, and body heat to heat the building. These buildings make use of heat energy that conventional buildings may exhaust outside. They may use heat recovery ventilation, hot water heat recycling, combined heat and power, and absorption chiller units.

The goal of green building and sustainable architecture is to use resources more efficiently and reduce a building's negative impact on the environment. Zero-energy buildings achieve one key green-building goal of completely or very significantly reducing energy use and greenhouse gas emissions for the life of the building. Zero-energy buildings may or may not be considered "green" in all areas, such as reducing waste and using recycled building materials, etc. However, zero-energy, or net-zero buildings do tend to have a much lower ecological impact over the life of the building compared with other 'green' buildings that require imported energy and/or fossil fuel to be habitable and meet the needs of occupants.

Because of the design challenges and sensitivity to a site that are required to efficiently meet the energy needs of a building and occupants with renewable energy (solar, wind, geothermal, etc.), designers must apply holistic design principles, and take advantage of the free naturally occurring assets available, such as passive solar orientation, natural ventilation, day lighting, thermal mass, and night time cooling.

The first consideration in a ZEB is to maximize efficiency in the building's energy demand or load. Designers should minimize the electricity load by utilizing integrated energy design strategies such as building envelope improvements, day

Fig. 3.4 Zero-energy house
using PV system

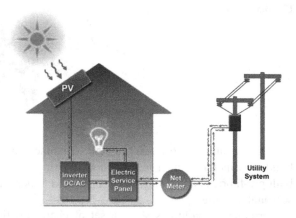

lighting techniques, and natural ventilation applications (refer to various energy design resources design briefs).

Additionally, installing energy-efficient lighting and cooling equipment throughout a building minimizes energy loads. The goal is to minimize the building's energy needs and then supplement the remaining loads supplied by the local utility grid with PV-generated electricity. By minimizing the electricity needs and utilizing for example BIPV, the designer maximizes the potential energy cost savings (Fig. 3.4).

Just as a building should be designed to maximize energy efficiency, another system such as a BIPV system should be designed to optimize electrical output. It is important to note that the availability of solar radiation generally matches commercial building electric loads throughout the day and throughout the year. For example, typical energy use for office buildings peaks near midday and during the summer season, the time when there is the greatest solar potential. For maximum energy output, it is important to determine the orientation, tilt angle, size and location of the BIPV system in relation to the building site and design. Flexibility exists in the placement (tilt and orientation) of BIPV, so it is best to match the time of day, month, and season when peak solar generation occurs with the peak electrical needs of the building.

Generally, there is no simpler way to save energy than turning off equipment when it is not needed that means scheduling equipment operation and "locking out" equipment operation when conditions warrant.

Needless operation after hours and on weekends is one of the largest energy wasters in commercial buildings. HVAC and lighting systems can be scheduled at the zone level, so that systems in unoccupied areas can be shut down.

Optimum start produces energy savings by starting equipment only as early as required to bring the building to set point at the time it will be occupied and optimum stop strategy determines the earliest possible time to turn off equipment before unoccupied periods begin and still maintain occupant comfort.

Fig. 3.5 Energy consumptions in European houses

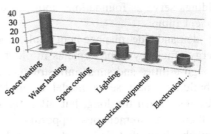

Residential Energy Consumption in Europe

Another technique is to bring the building to the desired temperature before occupancy after a night setup or setback with the least amount of energy, by closing outside air dampers.

In climates with a large nighttime temperature depression (dry climates), purging or flushing the building with cool outside air in the early morning hours can delay the need for cooling until later in the morning.

The boiler, the chiller, and associated pumps can be locked out above a set outside air temperature, by calendar date, or when building heating requirements are below a minimum.

In a ZEB it is needed to eliminate demand spikes by programing time delays between startup of major electrical load-generating equipment so that the startup peak loads stay below the peak demand later in the day. In general, there are two characteristics to reduce energy consumption and to make a ZEB:

- Clear and understandable using energy methods.
- Correct and describe a system that will actually perform its intended functions.

In an era of rising energy prices and increasing concerns about the adverse effects rising energy demand could have on the environment, there are tremendous opportunities in energy efficiency and conservation.

Efficiency and conservation represent vast and relatively untapped domestic energy resources capable of reducing the need for more costly production-side expansions in electricity generation. In transitioning to a new energy economy, the European Union can take the leading role by promoting energy efficiency to create even stronger economies, greener communities, and a healthier environment (Fig. 3.5).

Energy efficiency truly represents an undiscovered gold mine of energy and cost savings across Europe. If every country in Europe adopted energy efficiency best practices, the need for new generating facilities in the region could be reduced by as much as 75% over the next 15 years, the equivalent of 100 new large power-generating facilities. Project estimates a regional net economic gain of 37 billion euros, along with improved air quality, water savings, and reductions in regional greenhouse gas emissions.

Public and residential buildings offer energy-saving opportunities because they represent roughly one-third of all energy consumption in the EU. Homes and public buildings serve as foundations upon which the EU can begin to build an energy-efficient future.

Develop demand-side management tools, such as near-zero energy use in homes, energy efficiency improvements in older homes, energy-saving programs offered through utilities and smart infrastructure across the grid.

Public buildings must set a model example in energy-efficient and green building designs. Given the long, but often tight financing schedules for public buildings, there are tremendous opportunities to utilize performance-based financing mechanisms that fund and service buildings with long-term energy savings that accumulate throughout the lifetime of a building. They offer a win–win scenario: performance-based contracts pay for themselves and often achieve 25% or greater energy savings than conventional construction projects.

The greatest potential to generate energy efficiency gains:

- Improving energy efficiency education
- Strengthening code performance
- Rewards and incentives for energy-efficient practices
- Manage energy demand for efficiency results
- Decoupling and regulatory restructuring to remove utility impediments
- Innovative financing for public building projects.

Near-zero energy homes are built, operated, and maintained to achieve at least a 50% or greater improvement in energy performance over conventionally built homes through a combination of energy efficiency improvements and the use of on-site renewable energy systems, such as photovoltaic (PV) panels and solar thermal hot water systems to produce as much energy as the home consumes on an annual basis. They feature comfortable and traditional-looking home designs that perform well and require only standard maintenance.

Near-zero energy homes are growing in their potential to become more cost-effective for homeowners. PV panels often present a significantly higher initial cost by adding up to 20,000 euros to the price of a new home. However, near-zero homes have demonstrated energy savings that reduced homeowner utility bills by 60% or more, offsetting higher mortgage payments.

In many countries, tax credits and utility incentives are available to reduce the initial cost of efficiency features and renewable energy systems for builders and/or homeowners. To encourage greater investment and savings returns in near-zero homes, EU can offer compelling incentive options to stimulate investment in near-zero energy homebuilding.

Using smart infrastructure could be expanded to create a positive rate impact. In cases where utilities have the ability to charge time-of-use rates, many smart meters determine real-time price peaks, allowing residential customers to adjust their energy use to off-peak hours. An increasing number of utilities also offer voluntary "cycling" programs that deactivate electricity transmission to air-conditioning units by remote control as a way for customers to save energy during periods of peak energy demand without affecting home comfort. Time-of-use rate programs are encouraged to complement smart meter use.

The following tools and actions could encourage building occupants to capture greater energy savings from demand-side management programs:

- Employ greater use of smart infrastructure and metering to shape consumer awareness of electricity consumption.
- Expand smart infrastructure to serve as the price driver, charging time-of-use or off-peak pricing and installing new meters to have a rate impact.
- Allow demand-side management to be utilized to the fullest extent possible by including it in the rate base.
- Meet long-term demand needs by including DSM considerations as an integral component of energy portfolios.
- Encourage utilities to write comprehensive energy efficiency plans.
- Create a report card of utility EE best practices to determine what programs are working and collaboratively develop a matrix for effective program delivery and evaluation.
- Formulate a simple, efficient, and consistent method of tracking energy efficiency, such as quantifying DSM efforts in KWH per unit over time.
- Encourage utilities to put more energy efficiency information on utility bills to serve as important educational tools by including suggestions for reducing customer demand or statistics comparing electricity consumption to neighboring homes or normalized use.
- Include state energy efficiency goals as part of state renewable portfolio standards (RPS).
- Develop pilot programs within local governments that initiate and showcase construction efforts for near-zero homebuilding.

Construction of a ZEH involves many of the same materials and technologies familiar to the building trades and homeowners. Opportunities to reduce energy use exist in all areas of the home. The first opportunity to save energy is to reduce space heating and cooling and water heating loads.

This often means that more insulation is required, along with attention to other important features such as air infiltration moisture barriers, and ventilation. Major equipment in the home should also be of the highest efficiency that is affordable, and be sized and installed correctly. This includes the furnace, air-conditioner, and water heater as well as the duct and piping systems that deliver air and water to the outlets.

The next opportunity to reduce energy loads is to use higher efficiency lighting and appliances. The final opportunity is to be aware of energy use on a daily basis and turn off lights and appliances when not in use. Once the home's energy use requirements are reduced, a photovoltaic (PV) system is installed to provide the electricity used in the home and offset electricity supplied by the utility when averaged over the course of 1 year.

The means to achieving a successful ZEH are not the same for all homes. There are new building systems that can be used to increase the insulation of the walls and the roof, including structural insulated panels (SIPS), insulated concrete forms (ICF), and typical frame wall systems which can be used with new types of insulation.

Research into energy-efficient construction techniques will prove fruitful in designing and constructing the most energy efficient home for the least amount of money.

Some smart solutions to make a zero-energy building are:

1. *Decrease the energy requirements for space heating, cooling, and water heating:*

   - Orient the home with smaller walls facing west and include overhangs and porches
   - Increase foundation, wall, and ceiling insulation
   - Use low U-value, low-E windows in all climates and low solar heat gain (low SHGC) windows in cooling climates
   - Seal all holes, cracks, and penetrations through the floor, walls, and ceiling to unconditioned spaces
   - Install adequate ventilation, especially from kitchens and baths

2. *Increase the efficiency of the furnace (or heat pump), and the air-conditioner:*

   - Buy as high-efficiency equipment as affordable for the climate
   - Design the supply and return duct system appropriately and seal tightly using approved tapes or mastic
   - Consider ground-source heat pump technology where space and cost conditions permit
   - Where climate-appropriate consider alternative cooling systems such as ventilation only or evaporative coolers

3. *Install a solar hot water preheat system, an efficient backup water heater, and an efficient distribution system:*

   - Consider a parallel, small diameter piping system for the hot water outlets
   - Install low-flow fixtures
   - Choose water heating equipment with a high energy factor
   - Look for a knowledgeable solar hot water installation company
   - Evaluate solar systems

4. *Install efficient lighting fixtures:*

   - Consider permanent fluorescent fixtures in as many locations as possible
   - Look for the ENERGY STAR® label

5. *Install efficient appliances:*

   - Include the refrigerator, dishwasher, and laundry appliances
   - Look for the ENERGY STAR label
   - Compare appliance efficiencies

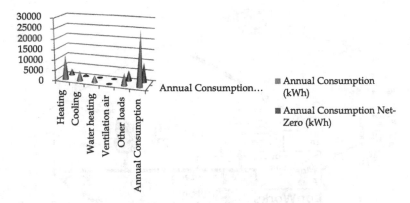

Fig. 3.6 Comparison of a zero-energy building and a typical construction one

6. *Install a properly sized photovoltaic system:*

- Look for a knowledgeable solar PV installation company
- Evaluate tax and other incentives
- Use PVWATTS for a quick estimate of PV output
- Find a certified solar PV installer

7. *Turn off lights, computers, and appliances when not in use*

The successful zero-energy home (ZEH) doesnot end with the designer and builder. The homeowner plays an extremely important role as they do with any well-maintained home. Throughout the life of the home, the homeowner has the most significant impact on the actual performance of the ZEH. Therefore, the ZEH homebuyer must be conscious of daily habits and patterns that affect energy use in the home as well as proper maintenance of equipment and appliances (Fig. 3.6).

For instance, understanding the way certain energy efficiency features of the home work such as programmable thermostats or photo-sensitive outdoor light fixtures are essential.

Simple things such as turning off lights when leaving a room or closing doors when performing even quick tasks outdoors can eliminate "wasted" energy. Paying careful attention to actual energy needs and avoiding unnecessary energy use are the first steps in ensuring that the ZEH performs as it was designed and built.

Secondly, as with any valued property, equipment in the home and the structure itself must be carefully maintained. Changing furnace filters, having heating and cooling systems cleaned regularly, periodically checking the operation of solar systems, and maintaining exterior caulking and painting are only a few examples of ways in which a homeowner can assure a long-lived, high-performance, and zero-energy home.

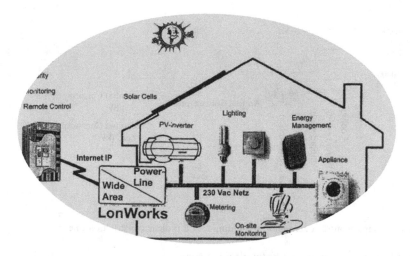

Fig. 3.7 Zero-energy smart house

A zero-energy home combines high levels of energy efficiency with renewable energy systems to annually return as much energy to the utility as it takes from the utility—resulting in a net-zero energy consumption for the home (Fig. 3.7).

Questions

1. What is a zero-energy building?
2. Give some simple solutions to make a zero-energy building?
3. Give some advantages and disadvantages of a ZEB?
4. How could encourage the homeowners to get a zero-energy house?
5. What are the characteristics of a zero-energy building?

Chapter 4
Solar Energy

Europe has a significant potential of wind, biomass, solar, and geothermal energy. Southern Europe with 300 approximately sunny and warm days per year, northern and islands' area with wind, an average wind speed exceeding 7.5 m/s and an important number of geothermal fields, Europe has all the criteria for wind, solar, and geothermal energy production. As the energy demand is projected to grow more and more over the next years, the biomass potential can cover a high percentage of its total energy consumption.

The development of Renewable Energy Sources has been among the major energy policy lines of European Union for the last 10 years. It is seen as an important contribution to the improvement of the European environmental indicators and, in particular, to the abatement of CO_2 emissions.

Legal and financial incentives are the tools of strategy to support renewable energy technology (RET) investments (Fig. 4.1).

The most favorable RES, especially in South European Countries is solar energy.

Solar electricity not only protects the environment, but it also creates independency from expensive energy imports and jobs. Ground-mounted PV installations are a cost-effective market segment and an integral part of that success (Fig. 4.2).

Daylight is solar energy. This is a trivial statement but comes lightly to the background when speaking of solar energy use.

Solar energy remains the largest source of potential, clean and renewable energy. Since the oil shock in the early 1970s the solar energy market has grown on average 25–30% per year. Although environment and energy security are the main reasons for the global trend of adopting alternative energy, however, the main incentive is the potential of this kind of energy to stimulate economical growth. Indeed, the economical impact of implementing and adopting alternative energy is highly considered by various stakeholders and lawmakers at both international and national levels.

E. V. M. Papadopoulou, *Energy Management in Buildings Using Photovoltaics*,   33
Green Energy and Technology, DOI: 10.1007/978-1-4471-2383-5_4,
© Springer-Verlag London Limited 2012

Fig. 4.1 Posts rammed into
the ground

Fig. 4.2 Solar module
frames made of aluminum

The main objectives of energy policy are:

- supply,
- utilization, and
- environmental objectives

Supply is based on reducing its independence from oil and by developing and utilizing alternative sources of energy.

Utilization is aimed at promoting and encouraging the efficient utilization of energy and discouraging wasteful and non-productive patterns of energy consumption within the given socio-cultural and economic parameters.

The environmental objective seeks to ensure that factors pertaining to the environment are not neglected in pursuing the supply and utilization objectives.

In fact, current or an expected user of PV needs detailed information allowing him prompt access to (i) the real received solar energy in his specific location and (ii) quantity produced of energy depending on the type of PV technology used.

So, all PV users need to know before buying or installing solar panel, if the location is suitable and receives optimal solar radiation, if the efficiency is enough for his daily demand of energy, and a monitoring and tracking tool such as a daily report about received and produced energy.

One of the most promising renewable energy technologies is photovoltaics. Photovoltaics (PV) is a truly elegant means of producing electricity on-site, directly from the sun, without concern for energy supply or environmental harm. These solid-state devices simply make electricity out of sunlight, silently with no maintenance, no pollution, and no depletion of materials.

Fig. 4.3 A PV skylight
entryway

Photovoltaic modules and solar collectors make the sun's energy usable, but technologies that provide for optimal light efficiency in buildings and that make "living and working with the sun" enjoyable also use solar energy.

The great advantage of photovoltaic (PV) systems is their highly modular structure. Building integrated PV-elements can easily be adapted to the architect ideas according to size, shape, and mounting technique. The generation potential of suitable mounting surface areas is very significantly.

Electronics, information technology, and PV-technology are all developing very rapidly. New ideas are implemented every day. Inverters and other electronic devices can now be built in large quantities, at lower cost, and in smaller sized units.

There is a growing consensus that distributed photovoltaic systems that provide electricity at the point of use will be the first to reach widespread commercialization. Chief among these distributed applications are PV power systems for individual buildings (Fig. 4.3).

Interest in the building integration of photovoltaics, where the PV elements actually become an integral part of the building, often serving as the exterior weather skin, is growing worldwide. PV specialists and innovative designers in Europe, Japan, and the US are now exploring creative ways of incorporating solar electricity into their work. A whole new vernacular of Solar Electric Architecture is beginning to emerge.

A Building Integrated Photovoltaics (BIPV) system consists of integrating photovoltaics modules into the building envelope, such as the roof or the façade. By simultaneously serving as building envelope material and power generator, BIPV systems can provide savings in materials and electricity costs, reduce use of fossil fuels and emission of ozone depleting gases, and add architectural interest to the building.

While the majority of BIPV systems are interfaced with the available utility grid, BIPV may also be used in stand-alone, off-grid systems. One of the benefits of grid-tied BIPV systems is that, with a cooperative utility policy, the storage system is essentially free. It is also 100% efficient and unlimited in capacity. Both the building owner and the utility benefit with grid-tied BIPV. The on-site production of solar electricity is typically greatest at or near the time of a building's and the utility's peak loads. The solar contribution reduces energy costs for the building owner while the exported solar electricity helps support the utility grid during the time of its greatest demand.

Fig. 4.4 Mono crystalline, Poli crystalline and Thin film A-Si:H

4.1 Photovoltaics (PV) Technologies

There are two basic commercial PV module technologies :

Thick crystal products include solar cells made from crystalline silicon either as single or poly-crystalline wafers and deliver about 10–12 watts per ft^2 of PV array (under full sun).

Thin-film products typically incorporate very thin layers of photovoltaicly active material placed on a glass superstrate or a metal substrate using vacuum-deposition manufacturing techniques similar to those employed in the coating of architectural glass. Presently, commercial thin-film materials deliver about 4–5 Watts per ft^2 of PV array area (under full sun). Thin-film technologies hold out the promise of lower costs due to much lower requirements for active materials and energy in their production when compared to thick-crystal products (Fig. 4.4).

4.2 Thin Film Panel Observations

- They are cheaper, but they need larger surfaces and structures
- The guaranteed output power is not as precise as in mono/polycrystalline modules
- There are no references from facilities producing an important amount of years.

A photovoltaic system is constructed by assembling a number of individual collectors called modules electrically and mechanically into an array (Fig. 4.5).

4.3 Building Integrated Photovoltaics (BIPV) System

Building Integrated Photovoltaics (BIPV) is the integration of photovoltaics (PV) into the building envelope. The PV modules serve the dual function of building skin—replacing conventional building envelope materials—and power generator. By avoiding the cost of conventional materials, the incremental cost of

Fig. 4.5 Types of cells

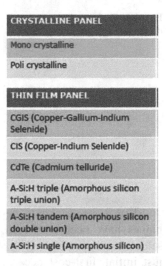

CRYSTALLINE PANEL
Mono crystalline
Poli crystalline

THIN FILM PANEL
CGIS (Copper-Gallium-Indium Selenide)
CIS (Copper-Indium Selenide)
CdTe (Cadmium telluride)
A-Si:H triple (Amorphous silicon triple union)
A-Si:H tandem (Amorphous silicon double union)
A-Si:H single (Amorphous silicon)

photovoltaics is reduced and its life cycle cost is improved. That is, BIPV systems often have lower overall costs than PV systems requiring separate and dedicated mounting systems.

A complete BIPV system includes:

- the PV modules (which might be thin-film or crystalline, transparent, semi-transparent, or opaque);
- a charge controller, to regulate the power into and out of the battery storage bank (in stand-alone systems);
- a power storage system, generally comprised of the utility grid in utility-interactive systems or, a number of batteries in stand-alone systems;
- power conversion equipment including an inverter to convert the PV modules DC output to AC compatible with the utility grid;
- backup power supplies such as diesel generators (optional-typically employed in stand-alone systems); and
- appropriate support and mounting hardware, wiring, and safety disconnects (Fig. 4.6).

BIPV systems can either be interfaced with the available utility grid or they may be designed as stand-alone, off-grid systems. The benefits of power production at the point of use include savings to the utility in the losses associated with transmission and distribution (known as 'grid support'), and savings to the consumer through lower electric bills because of peak saving (matching peak production with periods of peak demand). Moreover, buildings that produce power using renewable energy sources reduce the demands on traditional utility generators, often reducing the overall emissions of climate-change gasses.

Fig. 4.6 BIPV system
diagram

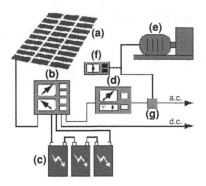

BIPV systems should be approached where energy conscious design techniques have been employed, and equipment and systems have been carefully selected and specified. They should be viewed in terms of life cycle cost, and not just initial, first-cost because the overall cost may be reduced by the avoided costs of the building materials and labor they replace. Design considerations for BIPV systems must include the building's use and electrical loads, its location and orientation, the appropriate building and safety codes, and the relevant utility issues and costs.

Carefully consider the application of energy-conscious design practices and/or energy-efficiency measures to reduce the energy requirements of the building. This will enhance comfort and save money while also enabling a given BIPV system to provide a greater percentage contribution to the load.

Choose Between a Utility-Interactive PV System and a Stand-alone PV System

The vast majority of BIPV systems will be tied to a utility grid, using the grid as storage and backup. The systems should be sized to meet the goals of the owner—typically defined by budget or space constraints; and, the inverter must be chosen with an understanding of the requirements of the utility.

For those 'stand-alone' systems powered by PV alone, the system, including storage, must be sized to meet the peak demand/lowest power production projections of the building. To avoid over sizing the PV/battery system for unusual or occasional peak loads, a backup generator is often used. This kind of system is sometimes referred to as a "PV-genset hybrid."

Shift the Peak

If the peak building loads do not match the peak power output of the PV array, it may be economically appropriate to incorporate batteries into certain grid-tied systems to offset the most expensive power demand periods. This system could also act as an uninterruptible power system (UPS).

Provide Adequate Ventilation

PV conversion efficiencies are reduced by elevated operating temperatures. This is truer with crystalline silicon PV cells than amorphous silicon thin-films. To improve conversion efficiency, allow appropriate ventilation behind the modules to dissipate heat.

Evaluate Using Hybrid PV-Solar Thermal Systems

As an option to optimize system efficiency, a designer may choose to capture and utilize the solar thermal resource developed through the heating of the modules. This can be attractive in cold climates for the preheating of incoming ventilation make-up air.

Consider Integrating Daylighting and Photovoltaic Collection

Using semi-transparent thin-film modules, or crystalline modules with custom-spaced cells between two layers of glass, designers may use PV to create unique daylighting features in façade, roofing, or skylight PV systems. The BIPV elements can also help to reduce unwanted cooling load and glare associated with large expanses of architectural glazing.

Incorporate PV Modules into Shading Devices

PV arrays conceived as "eyebrows" or awnings over view glass areas of a building can provide appropriate passive solar shading. When sunshades are considered as part of an integrated design approach, chiller capacity can often be smaller and perimeter cooling distribution reduced or even eliminated.

Design for the Local Climate and Environment

Designers should understand the impacts of the climate and environment on the array output. Cold, clear days will increase power production, while hot, overcast days will reduce array output;

Surfaces reflecting light onto the array (e.g., snow) will increase the array output;

Arrays must be designed for potential snow- and wind-loading conditions;

Properly angled arrays will shed snow loads relatively quickly; and,

Arrays in dry, dusty environments, or environments with heavy industrial or traffic (auto, airline) pollution will require washing to limit efficiency losses.

Address Site Planning and Orientation Issues

Early in the design phase, ensure that solar array will receive maximum exposure to the sun and will not be shaded by site obstructions such as nearby buildings or trees. It is particularly important that the system be completely unshaded during the peak solar collection period consisting of three hours on either side of solar noon. The impact of shading on a PV array has a much greater influence on the electrical harvest than the footprint of the shadow (Fig. 4.7).

Consider Array Orientation

Different array orientation can have a significant impact on the annual energy output of a system, with tilted arrays generating 50–70% more electricity than a vertical façade.

Reduce Building Envelope and Other On-site Loads

Minimize the loads experienced by the BIPV system. Employ daylighting, energy-efficient motors, and other peak reduction strategies whenever possible.

In addition, BIPV systems can be designed to blend with traditional building materials and designs, or they may be used to create a high-technology, future-oriented appearance. Semi-transparent arrays of spaced crystalline cells can provide diffuse, interior natural lighting. High profile systems can also signal a desire

Fig. 4.7 PV array

on the part of the owner to provide an environmentally conscious work environment.

Solar systems can be mounted on roofs, integrated into awnings or installed at ground level on rack structures. In general, roof-mounted systems are preferred because they require shorter runs, are less vulnerable to vandalism and are more aesthetically appealing than ground mounted systems. Where roof mounting is not an option, open, adjacent land is a potential alternative where security concerns are not an issue. Ground mounting does have advantages: easy access, high visibility, and easy expansion with additional panels.

To maximize the PV potential of a building, solar energy system design should be ideally considered and coordinated with the architectural design of a project. This provides the greatest flexibility in PV array configuration while ensuring that the installation is well integrated with the structure and end use.

South facing is best to maximize overall production, but it is still possible to achieve near-optimal production with a southeast or southwest site. Solar panels require essentially shade-free placement. The most common items that will cause shading are trees, other buildings, telecommunications, and HVAC equipment.

The best time to install a roof-mounted solar system is during construction or roof replacement to achieve the lowest installation cost.

PV panels have a 20–25-year warranty and solar hot water panels have a 10-year warranty and typically last longer. The roof must be strong enough to support the PV system. In general, PV equipment weighs about 3–5 pounds per sq. ft., depending on the technology used and installation methods.

Sites with year round and fairly constant hot water demand are most appropriate for thermal applications. Pools, dorms, prisons, and hospitals are good examples. Solar hot water systems will need a storage tank, which should be close to the existing hot water system to minimize pipe runs and heat losses.

PV systems should be as close to the electric meter as possible to minimize wire transmission losses.

4.4 Control Units

The PV installation and especially the inverter will become just another unit working together with all the electric equipment connected by a bus system. The PV electricity production could be monitored and communicated in more detail:

- PV Load Management and Information System
- house control system, security
- heating, cooling, shading, and ventilation control
- daylight-dependent control of electric room lights
- EIB-bus, building control system

The double use of building integrated PV-elements for shading and electricity production gives additional benefits to the house owner. The tilt angle of the tracking panels may be optimized for a maximum of electric output or optimal shading and light control of the office rooms. Therefore an air-conditioning system could be avoided. Such a system could be easily integrated in the central house control unit and information bus system. Together with other improvements such as low energy personal computers, PV will help to reduce the electricity demand of the building.

Central control units could switch on special appliances such as pumps or water heaters to optimize the in-house use of PV electricity during sunny periods. This would help the utility keep grid loads at lower levels. But it makes sense only if the tariff for the electricity exported to the grid is lower than the consumer tariff.

Questions

1. Give the main objective of PV policy?
2. How does BIPV system work?
3. Which principles should somebody follow while designing a PV system for building use?
4. What benefits a houseowner have by using a BIPV?
5. Notice the orientation issues of a PV installation?

Chapter 5
PV System in Buildings

The penetration of the photovoltaic systems in buildings increases rapidly. European Union and European governments have financed the installation of PV systems in buildings, which are connected to the grid, or not. For example, German government has supported a program to subsidise the installation of 100,000 grid connected PV systems in buildings. Furthermore, the Greek government has supported a program to subsidise the installation of PV systems in buildings by 70%. The estimated installed power was about 2 MWp.

On the other side, many buildings cannot be connected to the grid in an economic way even in highly industrialized countries such as Germany and Norway. As an example, more than 100,000 solar home systems (SHS) have been installed in Scandinavia on a commercial basis. These systems are mostly used to supply weekend-houses with electricity during the summer session. In the Alps, more than 30 PV-diesel hybrid systems have been set up to power alpine huts very reliably. The modularity and high reliability of PV modules are great assets that have to be used in the design of the systems technology. Modularization of components can lead to a significant cost reduction. Modular systems technology is based on coupling generators, such as PV, wind turbines, small hydro, and diesel gensets, and storage units on the established single- or three-phase AC grid. The communication compatibility of components may be achieved by using European Installation Bus (EIB).

The advantages of this solution are the availability of volume production of EIB components, high reliability, simple expandability as well as simple installation, use, and maintenance.

In this way, expandable hybrid systems can be constructed, forming AC grids of high quality and availability by transforming and storing different kinds of renewable resources. Different systems can be tailored to the specific supply conditions by selecting appropriate electricity generation units. Supply systems can be installed, with a few energy sources feeding in, by connecting them to the

E. V. M. Papadopoulou, *Energy Management in Buildings Using Photovoltaics*,
Green Energy and Technology, DOI: 10.1007/978-1-4471-2383-5_5,
© Springer-Verlag London Limited 2012

AC energy bus. The supervisory unit sets the Reference State of the power supply system and receives information about the actual state of each component.

Many companies, of the electric industry in Europe, have joined into only one technology for building Control components, which is EIB.

The use of EIB technology opens the way for compatibility of products from different manufacturers of PV components and Home Automation, opening the market in both directions. The electric loads of the house can be considered as loads of the Hybrid System. Some companies have produced EIB modules for Load Management, achieving the maximum energy saving. Such Load Management Systems for houses are currently available. They could control the household electric devices, during heating/cooling and lighting periods, in order to reduce the peak power consumption by 50% of the average prior state. Moreover, the energy consumption of these houses is expected to be reduced by 20–25%. This leads to a significant reduction of emissions by decreasing the fossil fuel consumption.

EIB technology is selected because different manufacturers produce a large variety of compatible modules to control the building appliances. EIB is compatible with other building automation systems and it was introduced as a communication bus to energy producing units, such as PV grid-connected inverters and Battery inverters.

EIB modules provide the opportunity to: control automatically and visualize the PV Hybrid and building systems.

There are no sufficient energy management installations in houses. Residents do not appreciate the importance of energy management because they do not know when they are exceeding the desirable energy consumption and how they are going to save energy. Owners, without thinking in long term, are interested in low cost installations when they are building their houses.

There are three main reasons to apply EIB:

- easy installation,
- increasing comfort, and
- energy savings.

Some scenarios that can be considered in order to achieve energy saving are

- If the outdoor brightness is sufficient for room illumination, decrease the light intensity without compromising the effectiveness of lighting. If a window or a door is open, turn off the heating/cooling.
- If nobody is present in the house, turn heating in the comfort temperature program.
- If the maximum desired level of the power consumption is exceeded, deactivate subordinate loads temporarily by the load management modules.

Energy Management System (EMS) for a complex system like a building has to control the components in accordance to the needs, especially comfort needs, of the occupants and it has to guarantee a minimized energy consumption. At the

Fig. 5.1 Typical PV hybrid building system structure using EIB

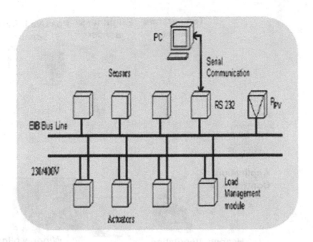

same time it should reflect to the user his actual consumption in a comprehensive way, a fact that it is ignored in conventional building management systems (BMS), where the occupant has no influence and no information at all due to the lack of interactivity.

The controller, based on EIB, includes an easy and intuitive dialog between user, energy customers, energy suppliers, and EMS as well as visualization of actual energy flows and costs.

Through EMS, PV Hybrid Systems can be operated in such a way that reliable power supply for all consumers becomes possible. The key element of the EMS is an information system in combination with variable pricing of the electricity. Under this type of EMS the actual electricity price is calculated according to the actual supply demand, which forms an energy market.

The actual price is also influenced by the actual state of charge of the battery, the electric contribution of the auxiliary generator, and the probabilistic future profiles for power production and demand. This system guarantees optimal use of PV energy because for periods of sunshine electricity is offered at a cheaper price. By contrast, at times of high power demand or during bad weather periods the electricity price will be raised, resulting in a reduced consumption (Fig. 5.1).

This system needs special metering, in which the actual power is being multiplied at each moment by the actual electricity price. Transmission of the actual price from the power plant (market) to consumers can be easily performed through the power line. With this way the EMS could convince the user, for example, to switch off the kitchen now and to postpone the cooking for some later time, depending the prediction for the load demand and the weather.

The European Installation Bus (EIB) is designed as a management system in the field of electrical installation for load switching, environmental control, and security, for different types of buildings.

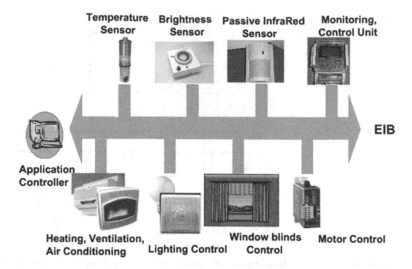

Temperature   Brightness   Passive InfraRed   Monitoring,
Sensor   Sensor   Sensor   Control Unit

EIB

Application
Controller

Heating, Ventilation,   Window blinds   Motor Control
Air Conditioning   Lighting Control   Control

Fig. 5.2 EIB equipments

Fig. 5.3 Conventional
electrical installation

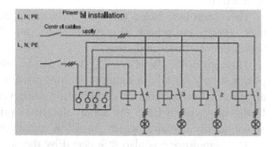

Its purpose is to ensure the monitoring and control of functions and processes such as lighting, window blinds, heating, ventilation, air-conditioning, load management, signaling, monitoring, and alarms.

The EIB system allows the bus devices to draw their power supply from the communication medium, such as Twisted Pair or Powerline (230 V mains). Other devices may, additionally, require power supply from the mains or other sources, as in the Radio Frequency and Infrared media (Fig. 5.2).

With EIB, all the required functions from any location in the building can carried out. It is also possible to operate the installation remotely, for example via a mobile or the Internet.

If several functions are to be executed using a single command, this can be implemented without problem. With central commands and user-defined procedures, all the shutters can, for example, be raised simultaneously, the constant lighting control activated and each room regulated to a separate temperature; all with a single push button action. Security functions can be integrated into the

Fig. 5.4 Electrical
installation with EIB

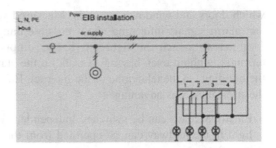

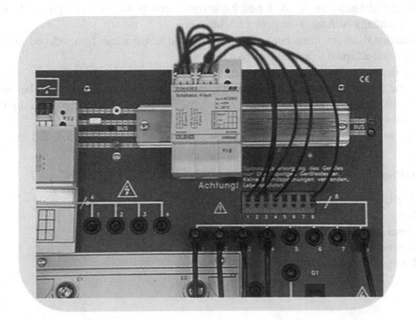

Fig. 5.5 Installation board

building installation. A security control panel manages all security-related signals and triggers alarms.

The security control panel can also be conveniently operated via EIB.

Signals can be displayed at any location or printed out via a logging printer.

In building installations using EIB, functions are not only executed via direct manual operation. Using "closed-loop" control systems, the user can preselect an individual daily profile for the room temperature or the room lighting level can be constantly regulated to a required value.

Time programs are recommended for regularly recurring events. Shutters and blinds can be raised automatically if the wind becomes too strong. EIB makes it possible for current information about the building installation to be displayed continually and can see at a glance in which rooms the lights are switched on or which doors and windows are open. Measured values can be shown on a display and alarm signals informing about possible dangers in building. If someone has forgotten to switch the light off, it can simply switch it off from the display terminal, without even having to walk up the stairs. It is also possible to monitor the building via the telephone or the Internet. Electrical installations with EIB offer the user numerous advantages

- Electrical loads can be switched independently of the electrical circuit (e.g. the light in the hallway can be operated from the lounge or from elsewhere in the house).
- Electrical loads can be switched by several sensors without complicated two-way circuits or remote-control switches.
- Functional associations between actuators and sensors can be modified at any time and adapted to individual requirements.
- All the functions can be programed so that they run automatically.
- Logic operations can also be created (e.g. if the brightness level drops below a specific value after 18:00, all the shutters are lowered and the light in the hallway is switched on).
- The switching states of electrical loads can be displayed (Figs. 5.3, 5.4, 5.5).

Questions

1. Are there any reasons to apply EIB?
2. What benefits an EIB usage gives to installation?
3. Describe some advantages of EIB?
4. Design a typical PV building installation with EIB?
5. Design a conventional electrical diagram?

Chapter 6
Photovoltaics Technology

Solar photovoltaic system or *Solar power system* is one of *renewable energy system* which uses PV modules to convert sunlight into electricity. The electricity generated can be either stored or used directly, fed back into grid line, or combined with one or more other electricity generators or more renewable energy source. Solar PV system is very reliable and clean source of electricity that can suit a wide range of applications such as residence, industry, agriculture ,and livestock (Fig. 6.1).

A building's photovoltaic power system includes solar panels, a grid-interactive current inverter, rooftop mountings and a full wiring system. Photovoltaics offer consumers the ability to generate electricity in a clean, quiet and reliable way. Photovoltaic systems are comprised of photovoltaic cells and devices that convert light energy directly into electricity. Because the source of light is usually the sun, they are often called solar cells. The word photovoltaic comes from "photo," meaning light, and "voltaic," which refers to producing electricity. Therefore, the photovoltaic process is "producing electricity directly from sunlight."

PV cells convert sunlight directly into electricity without creating any air or water pollution. PV cells are made of at least two layers of semiconductor material. One layer has a positive charge, the other negative.

When light enters the cell, some of the photons from the light are absorbed by the semiconductor atoms, freeing electrons from the cell's negative layer to flow through an external circuit and back into the positive layer. This flow of electrons produces electric current.

To increase their utility, dozens of individual PV cells are interconnected together in a sealed, weatherproof package called a module. When two modules are wired together in series, their voltage is doubled while the current stays constant. When two modules are wired in parallel, their current is doubled while the voltage stays constant. To achieve the desired voltage and current, modules are wired in series and parallel into what is called a PV array. The flexibility of the

E. V. M. Papadopoulou, *Energy Management in Buildings Using Photovoltaics*,
Green Energy and Technology, DOI: 10.1007/978-1-4471-2383-5_6,
© Springer-Verlag London Limited 2012

Fig. 6.1 Generating
electricity using PV

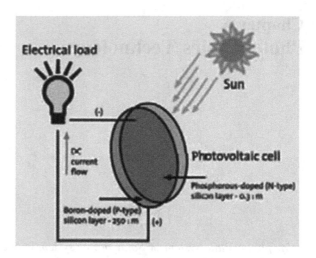

modular PV system allows designers to create solar power systems that can meet a wide variety of electrical needs, no matter how large or small.

For some applications where small amounts of electricity are required, like emergency call boxes, PV systems are often cost justified even when grid electricity is not very far away. When applications require larger amounts of electricity and are located away from existing power lines, photovoltaic systems in many cases offer the least expensive, most viable option.

In use today on street lights, gate openers, and other low power tasks, photovoltaics are gaining popularity around the world as their price declines and efficiency increases.

PV is one of the most environmentally friendly forms of renewable energy. Its many attractive features include the following:

- It has no moving parts;
- It is easy to install;
- It requires minimal maintenance;
- It consumes no fuel;
- It produces no pollution;
- It operates silently;
- The equipment has a long life span.

Benefits of using PV systems include:

- Using a pollution-free and infinitely renewable fuel source.
- Usable anywhere, where there is direct sunlight.
- Often less expensive than grid electricity hook-ups in remote places.
- Helping to create a more sustainable and independent energy future.

Table 6.1 Feed—in tariff

Year month	Interconnected system		Autonomous power systems	
	>100 kW	≤100 kW	>100 kW	≤100 kW
2009 February	400	450	450	500
2009 August	400	450	450	500
2010 February	400	450	450	500
2010 August	392.04	441.05	441.05	490.05
2011 February	372.83	419.43	419.43	466.03
2011 August	351.01	394.88	394.88	438.76
2012 February	333.81	375.53	375.53	417.26
2012 August	314.27	353.56	353.56	392.84
2013 February	298.87	336.23	336.23	373.59
2013 August	281.38	316.55	316.55	351.72
2014 February	268.94	302.56	302.56	336.18
2014 August	260.97	293.59	293.59	326.22
From 2015 and on	1.3 × ASMP	1.4 × ASMP	1.4 × ASMP	1.5 × ASMP

Anyone who has used a pocket calculator has seen PV at work. Light is a form of energy. It travels from light sources such as the sun in streams of energy units called photons. As photons of light hit the window above the calculator's buttons, they generate an electrical current in the silicon wafer that is in the window. The current is sufficient to power the calculator's computer chip.

6.1 Silicon PV Cells

Silicon, which is a basic constituent of sand and is one of the most plentiful substances on earth, reacts to light. When light shines on a silicon crystal, the photons strike the silicon atoms and dislodge electrons. The electrons form a current that can be conducted into electrical circuits and devices.

To augment the process, silicon PV cells are coated with boron, which lacks electrons and attracts them, and phosphorus, which has an excess of electrons and can provide an extra supply. A typical silicon PV cell, called a solar cell, consists of a layer of phosphorus-coated silicon in contact with a layer of boron-coated silicon. The cell, when connected in an electrical circuit and illuminated by sunlight, produces an electrical current which flows in the circuit. The stronger the light, the more electrons are dislodged and the larger the current that is produced.

Groups of PV cells are mounted together to form a *module*. Typical modules range in size from a few square inches to about 25 square feet. Groups of modules are mounted together to form an *array*.

6.2 Crystalline Silicon and Amorphous Silicon

Throughout its period of development, PV has utilized crystalline silicon as its electrical conversion material. In the most common manufacturing process, purified silicon is melted into a large crystal. Thin "slices" are cut, which are coated as described above. These become PV cells. With current technology, crystalline silicon PV cells can convert more than 12% of the sunlight that they receive, into electricity.

An alternative material called amorphous silicon is attracting increasing attention for home building. Amorphous silicon has no crystalline structure, and can be applied to substrates, including flexible substrates, in thin films. The conversion efficiency is generally lower than that of crystalline silicon, but it offers the potential of lower mass production cost.

The conversion efficiency of amorphous silicon was typically less than 6%. However, thin-film amorphous silicon cells that convert over 10% of sunlight to electricity are now days available.

Solar Photovoltaic Technology:

- Innovative manufacturing process technologies for raw materials, solar cell, and solar module production;
- Photovoltaic (PV) manufacturing equipment and automation expertise;
- High efficiency solar-LED (light-emitting diode) lighting systems;
- Novel photovoltaic module products;
- Hybrid concentrator PV-active solar thermal systems;
- Expertise in the design of solar electric systems for off-grid, remote, or northern locations;
- Design of solar electric systems for large-scale grid-tied applications;
- Consulting services for site assessment and load analysis;
- Portable solar chargers for consumer electronics;
- Industry leading solar charge controllers, inverters, rectifiers, and other balance of system components;
- Installation of off-grid power systems for rural telecommunication systems.

Photovoltaic systems can be installed on a variety of on- and off-grid applications, and can be integrated into buildings and other fixed structures. They have several inherent advantages over traditional power generation technologies.

- Energy from the sun is, for practical purposes, free, renewable, and inexhaustible;
- Solar electricity generation offsets greenhouse gas and pollutant emissions from fossil fuel generation, and toxic waste from nuclear generation;
- PV is peak power—it produces most of its energy when demand peaks (daytime) that is when power is typically most expensive to produce. Its increased use can also eliminate the need for new high-cost centralized power plants;
- PV can create increased autonomy through independence from the electricity grid or backup during power outages.

Fig. 6.2 SMA inverter
sunny central 500HE

Each connected building photovoltaic system (BAPV/BIPV—Building Applied/ Integrated Photovoltaics) can be analyzed in two individual building blocks: the solar modules, which convert solar energy into electricity and the electronic converter to perform the adaptation of electricity to requirements of low voltage. With the competitive prices of solar electricity we are demonstrating that there are no limits for photovoltaics (Fig. 6.2).

The PV industry expects solar PV electricity to become competitive with retail electricity prices within only a few years (grid parity).

The number of the needed Photovoltaic Panels (P/V Panels) determines the maximum produced power, while, the series or parallel connection of these determines the electrical characteristics (voltages and power) of the converters that are going to be used. Additionally, the proper functioning of the whole installation requires the use of certain auxiliary systems (Balance of System, B.O.S.), which guarantee both the safe connection of the inverter with the PV generators and electrical network and the resistance of the whole installation to mechanical stresses.

The electricity produced by photovoltaic panels is provided in the form of direct voltage and direct current. To make the supply of the electrical network of alternating current (AC) with the energy produced by the photovoltaics possible the mediation of appropriate electronic devices (inverters) is required. These electronic devices are known as electric converters, while, their part that has a task of connection to the electrical network and the conversion of the direct current to alternating current is known as inverter (Fig. 6.3).

Like every electrical installation of production and consumption of electric power that is connected to the alternating current network, in the same way, the electronic converters of the interconnected electric grid PV systems must fulfill the requirements specified by regulations and standards established or adopted by the operators of power systems and networks. Specifically, the connection of small dispersed power plants in the low voltage network is considered acceptable when the power supplied to the electrical network via electronic converters does not

Fig. 6.3 Module field and
inverter station

Fig. 6.4 PV roof mounted
and grid connected

affect in a negative way the quality of the power which is supplied to other
connected users (consumers or producers), does not disrupt the proper functioning
of parts responsible for the regulation and protection of the network, and does not
put persons and facilities at risk (Fig. 6.4).

An important distinction between electronic converters connected to PV
systems can be made depending on whether they include a transformer in one of
their scales. In the case of using a transformer, it may be of high frequency (Ferrite
Transformer) or of low frequency (iron-core transformer). The existence of a
transformer offers the advantage of galvanic isolation of the PV equipment from
the network of the alternating current. Although the low frequency transformers
lead to an increase of the volume and weight of the total construction, their
presence guarantees zero DC injection to the grid. Unlike the other topologies, any
asymmetry of the circuit power or of the control circuit can cause the appearance
of a small DC amount at the output of the inverter.

Fig. 6.5 Major photovoltaic system components

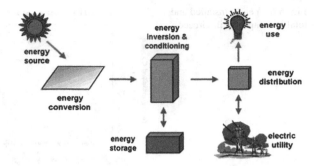

6.3 Major System Components

Solar PV system includes different components that should be selected according to building's system type, site location, and applications. The major components for solar PV system are solar charge controller, inverter, battery bank, auxiliary energy sources, and loads (appliances) (Fig. 6.5).

- *PV module*—converts sunlight into DC electricity.
- *Solar charge controller*—regulates the voltage and current coming from the PV panels going to battery and prevents battery overcharging and prolongs the battery life.
- *Inverter*—converts DC output of PV panels or wind turbine into a clean AC current for AC appliances or fed back into grid line.
- *Battery*—stores energy for supplying to electrical appliances when there is a demand.
- *Load*—is electrical appliances that are connected to solar PV system such as lights, radio, TV, computer, and refrigerator.
- *Auxiliary energy sources*—is diesel generator or other renewable energy sources.

6.4 Classification of the Building Interconnected PV Systems

Depending on the way the above structural units are combined, the connection to buildings PV systems of low power (up to 10 kW) are classified into two main technological trends. The Array technology and the Multi-array technology. The differentiation of the above-mentioned technological trends is depending on the number of the PV panels that are connected per electronic converter (power level of the converter) and on the way the PV panels are connected together (series, parallel connection, or combination of them).

Fig. 6.6 Yearly installed and
total PV capacity in Greece

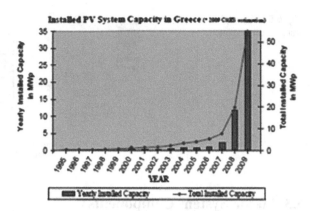

6.5 Building Connected PV Systems Under an Independent Producer

The Building Connected PV systems are under an Independent Producer (Feed in tariff). Therefore, all the energy produced by the power generating unit is sold to the power corporation and is not used for a partial or a total power supply of the building (self-consumption of the building) (Fig. 6.6).

6.6 Configuration of the Connection Depending on the Maximum Power of the PV System

The building PV systems of power up to 5 kWp are connected to the low voltage grid through a single-phase electric power supply as opposed to those whose maximum power exceeds the 5 kWp (But in no case up to 10 kWp). So, they are necessarily connected to the grid network through a three-phase electric power supply. In the case of the three-phase connection a symmetrical loading of the three-phases should be chosen Fig. 6.7).

It is mentioned that the percentage of asymmetry among the three phases of electric current must not exceed 20%.

6.7 Environmental Issues

PV power is generated without using depletable fuel resources, and the requirements for creating new or increased electrical transmission without generating PV reduces the pollution that is an inescapable feature of utility operation infrastructure.

Fig. 6.7 RV/Marine 12-V DC fluorescent lighting fixtures are good candidates for smaller off-grid applications

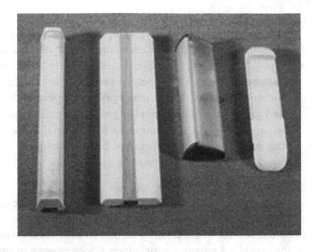

Photovoltaics are safe! It has far fewer risks and environmental impacts than conventional sources of energy. Nonetheless, there are some environmental, safety, and health (ES&H) challenges associated with making, using, and disposing of solar cells.

Fossil fuels produce acid rain, particulates, noxious fumes, carbon dioxide, and small amounts of heavy metals. PV, on the other hand, produces no pollutants during operation, making it a preferred option for offsetting emissions that result from fossil fuel use. In fact, 1 kW of PV could offset between 600 and 2,300 kg of CO_2 per year, as well as substantial amounts of other pollutants. And PV manufacturing produces only *modest* impacts, almost all from the energy needed to manufacture PV modules and systems. This energy is a problem only because it comes from conventional energy sources! Indeed, these initial energy costs of PV systems often can be paid back by PV-generated electricity in under 5% of a PV system's lifetime outdoors.

So, the industry is adopting technologies and procedures to minimize risks for each advanced PV option:

- *Amorphous Silicon.* Silane an– explosive gas– is used to make amorphous silicon. Toxic gases such as phosphine and diborane are used to electronically "dope" the material. To minimize explosion and toxicity risk, manufacturers use sophisticated gashandling systems.
- *Copper Indium Diselenide.* Toxic hydrogen selenide is sometimes used to make copper indium diselenide, a thin-film PV material. Manufacturers use gas handling systems to reduce risk, and use careful engineering and administrative controls to prevent exposure of workers or the public. Careful system design and gas detection systems can also effectively prevent exposure.
- *Cadmium Telluride.* Cadmium and its compounds, used to make cadmium telluride cells, can be toxic to lung at high levels of exposure. Inhalation of fine fumes or particles, more than ingestion or skin absorption, is the primary

concern. Manufacturers have effectively minimized exposure with engineering controls, personal protective equipment, and work practices. Biomonitoring of contaminant levels in workers is also a key defense against chronic toxicity.

- *Disposal and Recycling.* Because solar cells have useful lives of 20-30 years, waste generation will lag behind industry growth. Landfill leaching is a modest concern only because PV materials are largely encased in glass or plastic and many are insoluble. PV recycling will be challenging because of dispersed use, and small amounts of semiconductor material per cell. Machinery for dismantling modules for recycling has been developed, and recycling systems for batteries and electronics provide useful models.

Once a PV system has been purchased and installed, there is no cost for fuel. Utilities are permitted to pass increased fuel costs on to their customers. Utility costs are subject to uncertainty arising from natural disasters, increasing depletion of the world's fuel supply, or adverse occurrences in international affairs. A PV-equipped home escapes this future cost hazard for the portion of the electrical needs that is supplied by PV.

6.8 Feeding Electricity Back Into the Grid

Another factor relates to the capability of PV systems to generate power that can be fed back into the utility grid.

Solar panels on homes are typically installed using low-profile mounting brackets. The sleek panels are placed almost flush against the roof. Roof integrated solar shingles are available as well.

PV systems are known for their dependability; warranties are typically 20 years. A solar panel will generate a minimum of four times the amount of energy used in its production but there are panels which generate 20 times the energy required for their production.

The first step in considering solar is to adopt energy efficiency measures. A thorough energy audit will show homeowners ways to substantially reduce their home's energy use, and thereby reduce the size and cost of their PV system.

Key energy efficient measures that make any solar-powered home reasonable in cost are:

- Good air sealing and tight duct work.
- Energy efficient lighting.
- High efficiency appliances that meet or exceed ENERGY STAR® standards.
- Properly sized, high efficiency heating and cooling systems.
- Water heating provided by a solar water heating system.

The size of a PV system depends on the appliances, home size, and site. Proper solar panel installation requires a southern orientation that receives direct sunlight from 9 am to 3 pm daily.

PV systems may be designed to provide any amount of power depending on the specific needs of the homeowner. For example, a four-bedroom, three-story home shown using a 2 kW PV system could produce 30 percent of the energy needed to run the home.

A typical residential-size solar system installation will involve properly sized and installed AC and DC electrical wiring to reduce the risk of electrical fire, a proper grounding system to prevent shock and lightning damage, proper battery installation and venting to prevent gas explosions, and a properly installed solar array to maximize performance while avoiding roof damage.

The first decision to make is system size. For a few lights in two or three rooms, for example, the system requires one or two 12-V batteries. For more power as a small DC freezer or DC well pump, the system needs two to four 6-V batteries.

The size and number of solar modules needed depends on the capacity of the battery bank and where the off-grid dwelling will be located. The solar array must face south. A solar module produces the best year-round performance with a tilt angle equal to area latitude. A lesser angle will improve summer output, and a steeper angle will do better in winter. If the dwelling will only be used for part of the year, it should use the tilt angle that will produce more energy during that part of the year.

The solar array can be mounted on dwelling's roof, on a nearby pole, ground-mounted on a raised frame, or mounted on a nearby storage shed. Solar modules are fairly lightweight so main mounting concern is wind uplift, not caving in building's roof. Any mounting system should use only stainless-steel bolts or lag screws penetrating into rafters or blocking, since a strong wind will easily pull out any screws that only penetrate sheeting plywood.

Most of the 12-V solar modules sold today are smaller than 100 W. The current market trend is for larger modules, which require a nominal 24 V output. One deep-cycle 6-V golf-cart battery or 12-V RV battery will store approximately 1 kWh of electricity when discharged 50%. As a rough estimate, this means it will need about 200 W of solar array to recharge one battery in one day, assuming 5 h of direct sunlight.

This is typical for most summer months, but many parts of the northern European countries only receive three hours of direct sun during short winter days. This means it will need to increase the solar array to battery ratio, or simply reduce power usage during periods of cloudy weather.

There are many design problems associated with multiple "parallel" wired solar modules and batteries, so it is much easier to use larger-capacity solar modules and batteries than smaller ones. Ordering larger unit sizes will work out better in the long run than buying the smaller and cheaper brands typically found locally.

Watts for a given load do not change regardless of voltage, so two 100-W light bulbs that require 1.7 amps at 120 V AC (200 W/120 V), can be wired using a 14 size wire which has a 15 amp rating. However, at 12 V DC, this same load will draw 17 amps (200 W/12 V), which exceeds the 14 wire's 15 amp rating.

In addition, two 100-W light bulbs would only operate about 5 h before draining deep-cycle battery (5 h × 100 W × 2 bulbs = 1 kWh). This means it must to use only the most efficient DC lighting and DC appliances that can buy, and do not base wire sizing on 120-V AC loads.

Even using low voltage DC power system's installation needs fuses and circuit breakers. Each wire supplying a load must have a properly sized fuse or circuit breaker to prevent overload and possible fire.

Questions

1. Which are the major system components of a PV installation system?
2. Give some PV system's advantages over traditional power generation technologies?
3. What does a solar charge controller regulate?
4. What does it mean "feed in tariff" energy production?
5. What is B.O.S.?

Chapter 7
Installing PV System

Photovoltaics may be integrated into many different assemblies within a building envelope:

Solar cells can be incorporated into the façade of a building, complementing or replacing traditional view or spandrel glass. Often, these installations are vertical, reducing access to available solar resources, but the large surface area of buildings can help compensate for the reduced power (Fig. 7.1).

Photovoltaics may be incorporated into awnings and saw-tooth designs on a building façade. These increase access to direct sunlight while providing additional architectural benefits such as passive shading.

The use of PV in roofing systems can provide a direct replacement for batten and seam metal roofing and traditional three-tab asphalt shingles.

Using PV for skylight systems can be both an economical use of PV and an exciting design feature (Fig. 7.2).

An important issue for bringing PV into the home building mainstream is the successful integration of PV systems into home design. Solar collectors mounted on rooftops, sometimes at angles to the roofline in order to face the sun, have functioned successfully but have not always been regarded as improvements to neighborhood aesthetics.

A highly promising approach to integrate PV into home building involves roofing materials that incorporate modules. Usually, two types of flexible roofing/PV products that utilize amorphous silicon have developed, and are demonstrating and commercializing. One is designed as an exact substitute for asphalt roofing shingles and the other as an exact substitute for standing seam metal roofs.

For the modules, PV manufactures have developed a low-cost, roll-to-roll continuous manufacturing process for the amorphous silicon solar cell substrate, and a high-efficiency, thin film amorphous silicon alloy. These technologies offer relatively low material and process costs, and produce a lightweight, rugged, flexible substrate that lowers the installed cost of PV.

E. V. M. Papadopoulou, *Energy Management in Buildings Using Photovoltaics*,
Green Energy and Technology, DOI: 10.1007/978-1-4471-2383-5_7,
© Springer-Verlag London Limited 2012

Fig. 7.1 APS factory in
Fairfield, CA

Fig. 7.2 Intercultural center,
Georgetown University in
Washington, DC

Fig. 7.3 Installing PV tiles

The metal roofing module is an exact replacement for a standing seam metal roofing pan. An array utilizing these modules retains the roofing system, and integrates with the steel portion of any roof that utilizes this system. The PV modules can be installed by metal roofing tradespersons without the necessity for any additional or specialized training (Fig. 7.3).

Fig. 7.4 A typical domestic
PV installation

There are no electrical feed throughs or conduits on the roof. Each module has a pair of leads and a ground wire that the roofer permits to fall through the ridge vent at the top of the module during installation. The electrical hookup is performed independently by an electrician working in the attic area.

Other systems were built with products that feature innovative structural systems in home building, and approaches to achieve advanced residential energy efficiency (Fig. 7.4).

Some homeowners are turning to PV as a clean and reliable energy source even though it is often more expensive than power available from their electric utility. These homeowners can supplement their energy needs with electricity from their local utility when their PV system is not supplying enough energy (at nighttime and on cloudy days) and can export excess electricity back to their local utility when their PV system is generating more energy than is needed.

For locations that are "off the grid"—meaning they are far from, or do not use, existing power lines—PV systems can be used to power water pumps, electric fences or even an entire household.

While PV systems may require a substantial investment, they can be cheaper than paying the costs associated with extending the electric utility grid.

A grid-connected PV system will require a utility interactive DC to AC inverter. This device will convert the direct current (DC) electricity produced by the PV array into alternating current (AC) electricity typically required for loads such as radios, televisions and refrigerators. For an off-grid PV system, consumers should consider whether they want to use the direct current (DC) from the PV's or convert the power into alternating current (AC).

Appliances and lights for AC are much more common and are generally cheaper, but the conversion of DC power into AC can consume up to 20% of all the power produced by the PV system (Figs. 7.5, 7.6).

In addition to the array, components of a home PV system may include an *inverter* and *storage batteries.* These can be located in the basement or garage area of the home. PV arrays produce direct current (DC) rather than the alternating current ordinarily used in homes. The electricity generated by the array is fed into an inverter that converts the DC from the array to AC for the home, PV systems utilize storage batteries to store electricity generated at times of peak output, and make it available to the house during portions of the 24-h cycle when the array's output is low or has ceased.

Fig. 7.5 Basic solar cell
construction

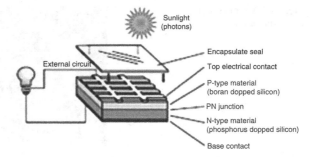

Fig. 7.6 Photovoltaic cells,
modules and arrays

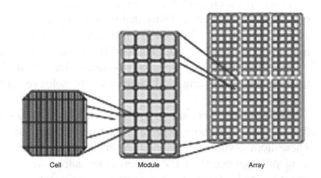

To store electricity from PV's, batteries will be needed. Batteries are often used in PV systems for the purpose of storing energy produced by the PV array during the day, and to supply it to electrical loads as needed (during the night and periods of cloudy weather). Other reasons batteries are used in PV systems are to operate the PV array near its maximum power point, to power electrical loads at stable voltages, and to supply surge currents to electrical loads and inverters. In most cases, a battery charge controller is used in these systems to protect the battery from overcharge and overdischarge.

The batteries used for PV systems are different from car batteries.

The batteries best suited for use with PV systems are called secondary or deep cycle batteries.

There are two types of deep cycle batteries: lead acid, which require the periodic addition of water, and captive electrolyte (or gel cell) batteries, which are maintenance free. In addition, PV systems require proper wiring, switches and fuses for safety, controllers to prevent the batteries from being overcharged or overly discharged, diodes to allow current to flow in the right direction, and grounding mechanisms to protect against lightning strikes (Fig. 7.7).

Considered in terms of wattage, a four-square-foot crystalline silicon module has a generating capacity of about 40 W. The output of a PV array varies in proportion to the available sunshine. The highest output, called the *peak wattage,* occurs at noon on sunny days.

Peak wattage and total electrical output are affected both by geography and by the season of the year.

Fig. 7.7 Components of
a typical off-grid PV system

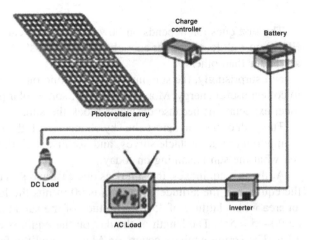

Amortization of the cost of PV equipment and its installation constitute the basic factors in the cost of PV-generated power. Equipment costs have been steadily declining. New products and technology are pushing the lower range of the cost down per kilowatt hour. When solar manufacturers and installers talk about costs, they usually speak in terms of euros per installed watt.

Simply put, PV systems are like any other electrical power generating systems, just the equipment used is different from that used for conventional electromechanical generating systems. However, the principles of operation and interfacing with other electrical systems remain the same, and are guided by a well-established body of electrical codes and standards.

Although a PV array produces power when exposed to sunlight, a number of other components are required to properly conduct, control, convert, distribute, and store the energy produced by the array.

Depending on the functional and operational requirements of the system, the specific components required may include major components such as a DC–AC power inverter, battery bank, system and battery controller, auxiliary energy sources and sometimes the specified electrical load (appliances). In addition, an assortment of balance of system (BOS) hardware, including wiring, over current, surge protection and disconnect devices, and other power processing equipment.

7.1 Orientation of the PV Panels

There are two basic questions to answer to determine if solar can work well at a location:

1. Does the location get enough sun?
2. Do nearby obstacles (trees, buildings, etc.,) at location block too much sun?

The first question depends on how cloudy the weather is. Within the USA and EU, almost all locations get enough sun to successfully use solar energy, but some are better than others.

Not surprisingly, the sun must actually shine on a solar collector in order for it to collect useful energy. More surprisingly, some solar projects are built that never meet expectations because obstacles block the sun.

The path of the sun across the sky changes with the time of year. This is why it is important to do obstacle survey, and not just stick the head out the window and see what the sun is shining on today.

At the two equinoxes, the sun rises due east and sets due west. At solar noon on the equinoxes, the altitude of the sun is 90 minus the local latitude. For example, an area with a latitude of 35°, the altitude of the sun at noon on the equinoxes will be 90–35 = 55°. The length of the day on the equinox everywhere on the earth is 12 h. The spring equinox occurs on Mar 21, and the fall equinox on Sept 21.

The winter solstice is the shortest day of the year and occurs on Dec 21 in the northern hemisphere. On this day the sun will rise well to the south of east, and will set well to the south of west. The altitude of the sun at solar noon will be 23.5° less than it was on the equinox— or, 55 – 23.5 = 31.5° in area example. This will be the lowest that the noon sun will be in the sky all year.

The summer solstice is the longest day of the year and occurs on June 21 in the northern hemisphere. On this day the sun will rise well to the north of east, and will set well to the north of west. The altitude of the sun at solar noon will be 23.5° more than it was on the equinox— or, 55 + 23.5 = 78.5° in area example. This will be the highest that the noon sun will be in the sky all year.

The 23.5° referred to above is the tilt of the earth axis of rotation relative to the plane of the earth's orbit. The summer solstice in the northern hemisphere occurs when the north pole is tilted toward the sun, and the winter solstice when the north pole is tilted away from the sun.

In planning a solar collector location, it is important to make sure that the sun will shine on the collector during all the parts of the year that want it to (Fig. 7.8).

7.2 Obstacle Survey

The obstacle survey checks for blockage of the sun by building, trees, hills, etc. It will need the following:

- A sun chart for the examine area
- A device to measure elevation and azimuth angles

The sun chart shows the position (azimuth and elevation) of the sun for every minute of the year.

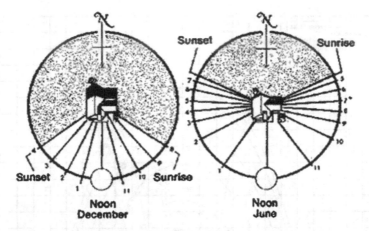

Fig. 7.8 Diagrams showing sunrise/sunset positions for the summer and winter solstices, and the suns altitude at the solstices and equinoxes

It will also need to make gages to measure the sun azimuth and elevation angles.

For most types of solar applications, it will be fine if exists 5 or 6 h of unblocked sun that is roughly centered around solar noon. If it has blockages during this 6 h period, it needs to evaluate how long they are, and at what times of year they occur (Fig. 7.9).

If blockages are serious, then must consider:

- Moving the collector to a better location,
- Trimming the obstacle (good for trees—not so good for buildings and mountains),
- Aiming the collector such that it gets more unobstructed sun—for example, if the sun is good in the morning, but blocked in part of the afternoon, then aiming the collector a bit east of south in order to get more morning sun.

Blockages caused by deciduous trees may be acceptable for solar applications like space heating, in that the trees will let most of the light through when the leaves are gone in the winter.

Photovoltaic panels can be particularly sensitive to partial shading. Shading even a small part of a panel can significantly cut the panels output.

Also, bear in mind that trees grow.

In order to achieve the maximization of the productivity of the PV modules, the best utilization of the solar radiation must be achieved. Since the direction of the sun changes as the time of day and the days of the year pass, it is known that in order for a panel to produce the maximum quantity of electrical energy it should be able to rotate so that it can follow the direction of the sun and be kept perpendicular to the direction of the solar radiation.

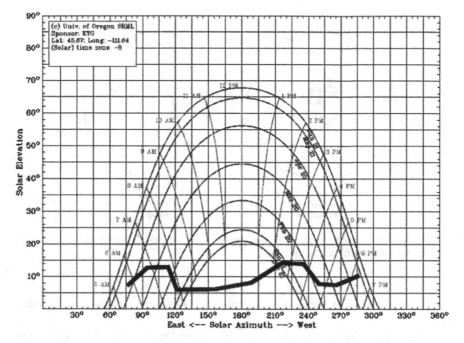

Fig. 7.9 Sun chart with obstacle horizon marked

Practically, the mechanical complexity and the cost of such a mechanism that allows the movement of the panels according to the way mentioned above makes it extremely difficult and costly to implement in building PV systems. So, the majority of building PV systems chooses an orientation of stable panels in order to achieve an average annual angle of 90° for the incoming solar radiation. This goal can be achieved with the right choice of the inclination and the right choice of the azimuth angle. The inclination of the module is expressed by the angle that is formed between the level of the surface of the PV panel and the horizontal level, while, the azimuth angle is formed on the horizontal level between the tilted side of the module and the local meridian with North–South (Fig. 7.10).

For the northern hemisphere, the best inclination of the PV panel for the maximum production throughout the year is equal to the geographic latitude of the place and the azimuth angle is almost 0° (with direction to the south).

Since, in the case of building PV systems the best inclination degrees and the orientation of the PV panel may be unattainable (because of restrictions resulting from the available spaces of the building), a solar radiation assessment of the surface where the PV system is going to be placed should be done. The reduction of the annual solar radiation (on the surface of the PV system) compared with the maximum theoretical value (best inclination and orientation degrees) is recommended not to exceed the 10% in order to maximize the economic benefits of an independent producer. Taking into account the constraints arising from the

Fig. 7.10 Orientation of the panels

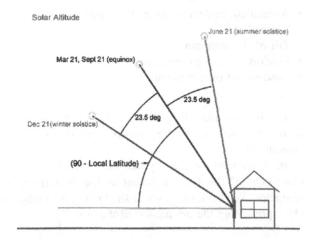

available surfaces of the buildings, it is generally preferable to have surfaces with south orientation deviation up to 70° to the direction of the south and inclination in the range of 0–50°. It is noted that inclination angles of more than 10–15° makes the self-cleaning of the panels from dust and other pollutants through the rain easier.

7.3 Shading Problems

Another important factor which affects the energy efficiency of a building PV system is the presence of shading. Taking into account that in a PV module, the PV elements (or parts of them) as well as the PV panels of an array, are connected together in series, it is obvious that even the shading of a part of the array can cause a significant reduction of the produced power compared to the expected value. Specifically, the total current power of a PV module array is determined by the reduced power supply of the shaded part of the PV series. Of course, if the shading limits the voltage of the shaded panels so low that the deviation route comes into conduction then the panel is excluded from the electricity production.

From another point of view, permanent or repeated local shading during hours of high solar radiation may stress the shaded PV module causing its premature aging. Therefore it is important to avoid shading even from objects of low volume like poles, antennas, or wires or even more from trees or nearby buildings, etc.

The choice of the placing position of the PV array should be done in such a way to ensure that there is no shading and that there will be no shading during the year and especially during the hours of the day with the highest solar radiation.

Depending on the type of installation, the shadowing study and the surface optimization, the project profitability may vary.

The main aspect to study are:

- Azimuthal deviation from the south (North hemisphere) or north (South hemisphere)
- Tilt of the solar panel
- Shadows of extern elements
- Shadows of own elements

Finally, to ensure long-term continuous performance of the PV system of any shadowing due to future construction of neighboring buildings should be considered.

In conclusion we can say that the general rule of choosing a location for the placement of the PV equipment is the horizon to the south to be free from obstacles. To check for possible shadowing throughout the year it is good to use a chart presenting the orientation of the sun.

7.4 Static Study Materials and Support Material

The placement of the PV modules in the building can be implemented either on an additional metal structure, or on the surface of the roof, or even with the integration of the modules to structural shell of the building. Although the PV array and the support base weight is unlikely to affect the static strength of the building, when the panels are installed on rooftops or roofs it is preferable to make a static research first (Or even a special study where required) so as to analyze the mechanical stress and the wind pressure of the surface where the panels are placed.

The PV modules are placed in a supporting system and ensure the proper function and the safety of the structure in extreme wind conditions, snowfall, ground quake, and temperature variations. These extreme weather conditions or a combination of them as well as the safety coefficients are determined by Eurocodes together with additional checks exactly like it happens in every construction test. For the efficient static of the supporting system it can be asked from the supplier to provide the relevant certificate. The support system can be part of a glazing panel or a joint of a roof or to be an independent system placed in such a way that creates a sun shade. The supporting system can be even metallic made of aluminium or heated galvanized steel or made of plastic (especially in the case of support areas). There is a variety of supporting systems available in the market. In each case it must be paid attention if there is compatibility with the rest elements of the equipment, furthermore in the accuracy of the certificates of the static efficiency of the entire construction. The way the PV modules are tighten must be in accordance with the specification standards of the PV module and additionally the dimensions of the panel must be relevant (or smaller) compared to those that have been approved in the static study for the issue of the certificate of static efficiency. As it concerns the connection of the supporting system to the building especially when it is placed on a rooftop the proper support must be used. This can be achieved by adding extra weight or with the use of screws. In the first case, the extra weight that is going to be placed must be in accordance with the static study of the building. In the case of using screws, they must not damage the existent insulation. In both cases, like in the case of another system, the specification standards for the supporting procedure are provided by the supplier of the supporting system. However, the compatibility with the building should be checked by a mechanic. Finally, the installer should be aware of the diversity of the support systems and their advantages and disadvantages including the ease of installation, reliability, and operational elements (such as the possibility or not of a natural ventilation of the module).

7.5 Placement Area of the Electronic Converters

One of the issues that deserve attention when designing a building PV system is the choice of the placement area for the electronic converters. Usually, the converters of the power generating units are placed either inside the buildings or in a specially designed indoor space which can be close to the PV equipment. Indeed, if the latter case happens there is a significant reduction of the length of electrical conductors of direct current (DC). With the direct result of reducing electrical losses, voltage drop, and reduction of wiring cost.

Of course there are electronic converters which, according to the manufacturer's technical manuals, can be installed either under the PV panels or in their supporting mechanism if there is enough space. Specifically, in the manufacturer's manual the degree of the protection factor (IP) of the converter from dust and water, as well as the temperature limits within which the safe and smooth operation is not affected should be sought. Otherwise, the adoption of this type of installation can cause a reduction in the life expectancy of the converter. Also, taking into account the fact that the electronic converter cooling is strongly influenced by the climatic conditions of the region in which the PV system is installed (weather temperature, conditions of sunlight, humidity, and wind), it is understood that when the converter is placed in an indoor space, near the PV equipment may be necessary to place a cooling mechanism (fans).

Questions

1. What will require *a* grid-connected PV system?
2. How many types of deep cycle batteries are there?
3. What does an obstacle survey check?
4. What kind of Shading problem should be avoided in a PV installation?
5. Give the main issues of Static Study Materials and Support material?

Chapter 8
Designing the PV System

The proper design of a PV system and the proper installation are required to ensure the smooth operation of power plants, both in terms of safety and energy efficiency.

8.1 Typical Electrical Values of a PV System

The maximum voltage expected from a PV array is the total voltage of an open circuit of panels connected in series, regarding the lower expected operating temperature.

The maximum expected value of a PV array current is calculated from the current value of a panel multiplied by the factor of 1.25. In parallel channels, the maximum expected value of the total current arises by multiplying the value of one channel with the number of the parallel channels. The safety factor of 1.25 covers specific atmospheric conditions and reflections that can occur in clear skies after rain (Solar irradiance <1,000 W/m^2).

The value of electricity calculated in this way should be taken into account in the sizing of cables and protections.

8.2 Temperature

The maximum expected operating temperature of PV modules and connecting boxes can reach the 70°C, in structures that allow the free circulation of air in the back side of the PV modules. In cases where the free circulation of air is prevented higher temperatures up to 80–90°C are expected. In the case where the interconnection conductors of the PV modules are very close to the PV module, their

E. V. M. Papadopoulou, *Energy Management in Buildings Using Photovoltaics*,   73
Green Energy and Technology, DOI: 10.1007/978-1-4471-2383-5_8,
© Springer-Verlag London Limited 2012

Fig. 8.1 Multi power stages

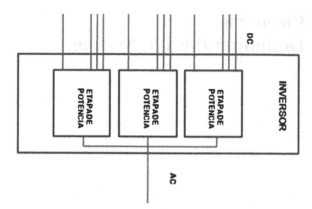

Fig. 8.2 Multi controlled

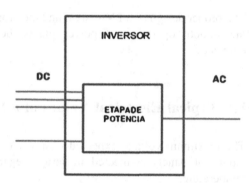

temperature should be seriously taken into account both for the correct choice of insulation of the conductors, as well as for the appropriate choice of their cross section (select the proper growth section corrective rate).

8.3 Cooperation Between the PV System and the Inverter

The inverter can be considered as the heart of a solar facility. Its cost, in relation to the complete installation, is between 6 and 9% but its performance is already between 95–97% .

So, it is important to know about their operation principles. We can find three options:

- Multi power stages one power stage (Fig. 8.1)
- Multi controlled (Fig. 8.2)
- Power stages (Fig. 8.3)

Fig. 8.3 Power stages

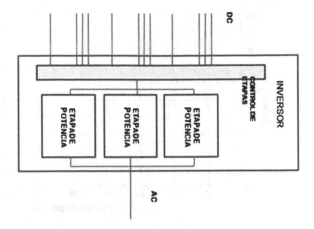

Fig. 8.4 I–V curve

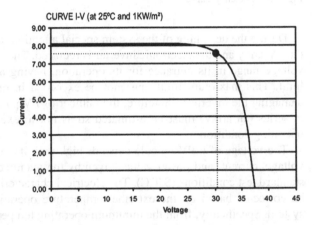

The inverter main features are:

- Maximum input voltage:

  The PV generator voltage must be under the inverter maximum input voltage

- MPPT voltage:

It is the range where the inverter is able to get the maximum power point from the PV generator I–V profile. The PV generator voltage must be within this range in the different conditions and weather during the whole year (Fig. 8.4).

The inverter has a different efficiency depending on the load. Usually, the manufacturers give the maximum efficiency and the european efficiency, which is the weighting of the different efficiencies when the load is: 5, 10, 30…100% The inverter temperature range is really important, as in some places the temperature can reach over 40°, and extra cooling might be considered (Fig. 8.5).

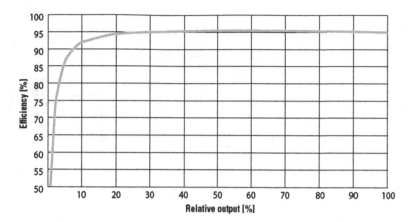

Fig. 8.5 Efficiency curve

During the designing of the system special attention to the cooperation between the PV array and the electronic inverter is required. The inverter requires a specific voltage range in its entrance for its operation, having a maximum input voltage limit. This maximum limit must not be exceeded in order to avoid any risk of damaging the inverter. Therefore, the number of PV panels that can be connected in series (an array) must be estimated so as not to exceed these limits in every operating condition.

The voltage of a PV module depends highly on its operating temperature. The voltage, current, and power values given by the manufacturer is mentioned to the standard test conditions (S.T.C). The electrical characteristics of PV modules must be corrected based on the extreme temperature operating conditions of the PV system. Specifically, from the minimum operating temperature of the modules, the maximum voltage value of the chains is calculated, and from the maximum operating temperature of the modules, the maximum value of the current of the parallel chains (branches) is calculated.

The maximum number of PV modules in series is calculated in a way that the total voltage of the open circuit of the array, within the lower operating temperature, must not exceed the maximum input voltage of the inverter. For example the lowlands of Greece the minimum temperature can be determined from −5°C or −10° C. (operating temperature of the active material of PV module). At the same time, the maximum operating voltage of the PV module, which must be greater than the open circuit voltage of the array within the lower expected operating temperature, must be checked in order to avoid a problem in the isolation of the PV module.

The minimum number of PV modules in series is determined so that the total voltage for an optimal functioning of the array to the maximum expected operating temperatures must not exceed the minimum input voltage of the inverter in order to be activated.

If the manufacturer provides only the value of the thermal coefficient for the voltage of an open circuit (V/° C), then the same value can be used for the voltage of the maximum produced power of the PV module without significant errors.

The minimum number of PV modules in series is determined so that the total voltage of excellent performance in the maximum expected working temperature to exceed the minimum voltage of the inverter's entrance width in order for it to be activated.

If the manufacturer only provides the value of the thermal coefficient for the open circuit voltage (V/°C), then the same value can be used for the voltage in the point of maximum produced power of the PV module, without any significant errors.

If from the series connection of the PV modules a power approximately as the nominal power of the inverter is not produced, then more parallel branches (acceptable number of series modules) must be connected so that the total power of the PV module array approaches the one of the inverter.

The operating current of the parallel branches must be lower than the maximum entrance current limit of the inverter. The total power of the PV array may exceed the nominal power of the inverter.

Finally, one important issue that must be taken into account is the compatibility between the type of the PV modules and the inverter which is relevant with the demand or not for grounding of the array in the side of the DC current. To be more specific, some types of PV modules according to the manufacturer demand grounding either of the negative (thin-film) or of the positive (back contact) pole.

The grounding must be either directly made, either through big resistance aiming to avoid functional problems that the above-mentioned types of PV modules have when they are not grounded (problems of corrosion and performance demission). Therefore, in such cases the use of inverters without galvanized insulation must be avoided, due to occurrence of leak currents, unless it is certified by the inverter's manufacturer that the chosen type of the inverter is appropriate for use with the modules we have chosen.

8.4 System Installation

For the full completion of the power generation plant the rules of the international experience and the existing regulations must be followed in order to avoid situations where human lives could be put at risk or material damages could be caused.

In detail, to complete the installation from the side of the AC current, the regulations deriving from the HD384 regulation must be followed. It is noted that the PV modules have different properties than those of the compatible sources. Those characteristic features come from the nature of the construction materials of PV modules and must be seriously taken into account in order to correctly design and complete a PV system.

In detail:

Taken into account the nature of the PV modules, it is proven that the PV modules act as electricity sources controlled by voltage. In fact, the maximum value of the current of a PV module is slightly bigger than the value of the nominal current of the module. Therefore, the use of safety fuses does not guarantee the pause of the system in case of error (module short-circuit). A short-circuit error on the side of the DC current may still exist despite the use of safety fuses, except in the case where the PV system is consisted of more than three parallel module series. In such a PV system structure the safety fuses may be used to protect each module series separately.

Unlike most power generation plants where the electricity production may be interrupted with the help of a general decoupling medium, the PV modules produce voltage on their edges as soon as they are exposed to sun light. Consequently, the installation of a PV module system is completed under voltage circumstances from the side of the modules.

During the completion of the electrical installation from the side of the DC current undesired situations may come up when:

1. Bad or loose connections exist formation of an electrical arc)
2. Error regarding the grounding (insulation damage and contact of active pipe with grounded metallic module or its support equipment)
3. Short-circuit error (insulation error and contact of active pipes)

8.4.1 Error Investigation

If an error regarding the ground occurs in a parallel branch, then the current of all the other branches will supply the error, creating an alternative current in the modules of the branch with the error.

A similar situation occurs when short circuit fault is created by the parallel branch or when the negative pole is grounded unintentionally by a first error in the ground that has emerged, followed by the second error in the ground.

The current error is fed by the PV modules and can remain even if the PV array is isolated from the converter, without having to interrupt the loop of the current error. This current can destroy the wires and the PV modules.

You can deal with the problem in the following ways:

1. With the dimensions determination of any parallel conductors' branch to withstand the current of N-1 parallel branches, as long as the current is lower than the maximum allowed AC current of the PV module. For most of the commercial PV modules the maximum allowed current may be considered the one equal to three times the short circuit current value.
2. With the installation of safety fuses on each side (positive or/and negative at the same time according to the location of the inverter) of each parallel branch.

The use of diode-return can solve the problem mentioned above, but it will burden the energy performance of the power generation plant because of power losses.

8.4.2 Protection

Regarding protection issues, the method of automatic feed disruption is not possible because of the special features of the PV modules. To protect them against direct or indirect contact, the use of very low voltage is possible (SELV or PELV systems). A PV module system is characterized as very low voltage system when the voltage of an open circuit in standard test conditions does not exceed 120 VDC. This case though is special and of low interest since most market products operate in higher voltages. Taking into account the fact that the PV modules that are used must be Class II regarding the insulation (according to standards EN 61730, in the application category—Application Class A—of continuous voltage of the system over 120 V), the insulation of the PV modules of this category must be insulated by Class II. The recommended practice for protection against indirect contact is the minimization of the possibility of error occurrence, beyond the use of PV modules Class II, and with materials and installation practices that ensure protection Class II or equal ("ground fault and short circuit proof installation"). The protection with material of category II, with enhanced insulation, is based on the fact that it is so strong that practically cannot be destroyed. The standard IEC EN 61730 consists of two parts. The first is about the minimum standards regarding the good construction of the PV modules, including the electricity insulation, for facilities where the maximum AC voltage may come up to 1,000 V. As it concerns the insulation, the standards of Class II must be fulfilled. The second part of the IEC EN 61730 refers to the test standards of the PV modules.

To protect the installation against overvoltage we must install high energy varistors close to the element that we want to protect.

The main aim of this device is to detect an overvoltage within a certain period of time and then divert it to the ground. *The device may be destroyed depending on the power to be diverted to the ground* (Fig. 8.6).

8.5 Grounding of the Inverter

The grounding (direct or neutral, depending on the area) aims at the protection of the production facilities and the safety of the individuals and must be performed according to the relevant regulations (HD384). At this point it must be noted that the grounding of one of the edges of the inverter on the side of the DC current is not compulsory in the European countries, unlike the USA.

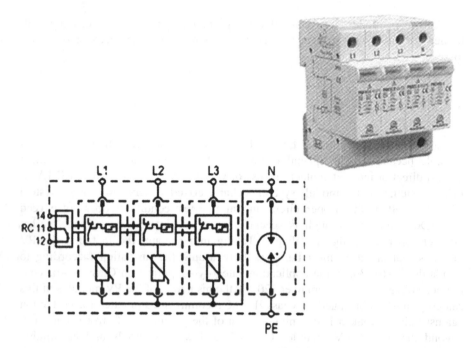

Fig. 8.6 Overvoltage protection

It is pointed out that the grounding or not of the DC current side depends on the PV modules technology and the topology of the inverter. The PV arrays that are formed by specific kinds of modules (thin film, back-contact), are grounded according to the manufacturer's recommendation so as to ensure their unobstructed function as well as their maximum performance. To be more specific, the PV modules of thin film with materials like a—Si and CdTe, due to their manufacturing technology (superstrate technology), usually show high danger of corrosion of the TCO layer, fact that causes damaging affects. To avoid such incidence, the negative edge of the PV source is grounded. This does not appear, according to studies, in thin film PV modules with other materials (e.g. CIS). In PV module systems with back-contact technology, it is compulsory (by the manufacturer) the grounding of the positive edge on the side of AC current so as to optimize their performance. In this case, the grounding must be done through high resistance. Under these circumstances, the use of inverter with converter is compulsory, unless it is certified by the manufacturer that the chosen type of the inverter (without insulation converter) is suitable for the modules that we have chosen. As it concerns the usual crystal modules, no specific demands have been put by the manufacturers regarding the grounding (or not) of the edge of the DC current side. In such cases, since active parts of the DC current side are not grounded, the use of inverter without isolation converter is possible. At anytime, the studier of the system must follow the instructions of the manufacturer

Fig. 8.7 Earthing system

regarding the demands that arise depending on the technology of the PV modules. Usually, the inverters' manufacturers, taking the above mentioned into account, suggest the suitable equipment according to the module type. It is stressed that, if the inverter does not include isolation transformer, the DC current side is not grounded. On the other hand, other exposed metallic parts of the PV module equipment (e.g. support base and metallic parts of the PV modules) must be grounded (Fig. 8.7).

Generally the typical elements used in every electrical installation are:

- Earth peg: different sizes depending on the required depth (from 1.5 to 2.5 m)
- Cable: copper without cover >35 mm^2

For low-power installations, it would be enough to use several earth pegs connected by a copper cable without cover.

For high-power installations, a copper cable grid is usually used without cover. Depending on the physical measures, earth pegs can be also used.

Finally, it is stressed that the grounding of the PV equipment may cause serious problems in case the insulation of the modules is not proper. On the other hand, although the use of ungrounded PV modules limits the above-mentioned risk, it increases the risk of damage of the modules caused by a potential lighting strike. In some cases the grounding of the modules is mandatory according to the manufacturer.

8.6 Wiring

On the side of the P/V the design and implementation of the wiring installation should provide protection equivalent to insulation of Class II.

Fig. 8.8 Groundloop in
trench

The wiring includes connections between the PV panels, the connections from
the edges of each P/V module of the array to the parallelism box, if it is used, as
well as the connections from the edges of the P/V array, for example the parallelism box, to the inverter.

All the cables that are exposed to the solar radiation should be resistant to UV
radiation (excluding common wires with PVC insulation).

Cables that are used in the modules' connection should have resistant insulation
for at least 70° or higher if there is not free circulation of air.

Choosing the right type of cables is important for the safety and the longevity of
the installation as well as for the satisfaction of the requirement for insulation
equivalent to Class II. For the connections between the modules, flexible single
core cables which fulfill the minimum requirements mentioned above with
increased insulation are usually used. The combination of these requirements is
difficult to be satisfied by using common wiring and requires the use of special
blends of plastics for insulation.

The cables can be aerial, but must be supported so as not to strain the connections. The support is achieved with the use of materials resistant to ultraviolet
rays, moisture, high temperatures, and corrosion (Fig. 8.8).

The P/V modules must have (bypass diodes), for alleviating the effects of
shadowing. For the connections between the cables it is recommended to use the
appropriate special connectors for a fast connection. The pre-installed cables of PV
panels must not be removed or replaced by cables of other type or cross section.
The routing of cables from the parallelism box to the inverter should ensure
protection equivalent to Class II. The cables must be of single core with double or
reinforced insulation. Otherwise, they should be placed in different channels. In
the connection boxes different areas with insulating separator must be used for
connecting the negative and positive conductors. Alternatively separate boxes for

positive and negative conductors can be used. The boxes to be used should be insulated and opening with a special key or tool.

The cross section of a cable is determined by the maximum expected current of a branch. The correction due to the temperature, which can reach the 70° C for the wires near the PV modules, should be taken into account.

It is noted that in a temperature of 70° the correction factor for wiring insulation resistant up to 90° must be 0.58. In this case, the cross section of the cable must be calculated based on the expected value of the maximum current multiplied by 1.72 (= 1/0.58) in order to not exceed the limits of the insulation resistance. Another factor that must be taken into account in order to determine the right dimensions of the cable is the energy losses. It is generally considered that the energy loss through the length of the DC cables based on the nominal values must not exceed the 1% of the nominal value of the PV system. This criterion usually leads to the choice of larger cross section.

In the side of the DC current a switch (when it is not included to the inverter) which is going to isolate the inverter from the P/V array should be installed. This switch should have the ability to isolate the loaded inverter (thus, the fast connectors do not fulfill this requirement as a means of isolation). The switch must be designed in order to work for DC current so as to isolate the two cores (ungrounded system).

8.6.1 AC Current Side

General rules must be followed on the side of the AC current.

The output of the inverter is connected to a different electrical board, where the protection and control measures are installed. The feed of the electrical board must come directly from the supply that the network's administrator has provided to the building.

The electronic inverters must provide isolation ability of their output from the AC current network.

The installation of the escape relay to the output of the inverter (AC current side) is made according to the demands of the standards. To be more specific, in case the inverter does not include galvanic insulation or includes high frequency converter, protection through escape relay type B (according to the IEC 364-7-712 standard) must be provided. The chosen inverter is best to already have this option without the need to install extra electrical equipment. The inverters of this type may have a certificate of measurements for the non infusion of DC current; therefore escape relay type A may be installed. For the selection of Isn current, besides the demands of the standards, it must be taken into account the fact that in PV installations with inverter without converter a leak current exists in the normal function of the system, the value of which cannot be accurately predicted (depending on the type of the modules, the inverter, and the weather conditions).

In those cases, the installation of an escape relay with stimulation current 30 mA may cause undesirable pauses to the function of the PV system.

It must be stressed that the minimization of the routing is undesirable, on the side of the DC current as well as on the AC current in order to achieve reduction of electrical loses.

8.6.2 Markings

Warning signs that all active parts inside the boxes remain active even after the isolation of the PV modules from the converter must exist in all connection boxes. The signs must be resilient to the environment where they are installed.

8.7 Building Protection of PV Modules Against High Voltage and Lightning Strikes

The protection of building PV systems against high voltages and lightning strikes is an issue that aims to protect the production facilities but mostly to protect humans.

With every precaution, based on the existing experience of hundreds of PV systems smaller than 10 kWp that were installed in European countries and do not project significantly from the building, the risk of direct lightning strike does not increase. Therefore, it is recommended to evaluate the risks of lightning strikes and the high voltages they create in order to ensure the safety of humans and building electrical facilities.

It is recommended that the facilities must avoid forming large current loops because a possible lightning strike will lead to the appearance of induction high voltages. If the PV system is installed in a building that already has lightning protection and a safety distance (0.5–1 m) can exist between the PV system and the conductors that collect and descent the lightning current, then the PV system is considered being within the protection area of the lightning rod must not be connected conductively to the lightning protection system (as long as it is about integration in below existing buildings). If a safety distance cannot be maintained then a conductive connection with the conductors of the lightning protection system is necessary.

A lightning may produce a *transitory overvoltage* of short duration, with a huge amplitude. The overvoltage produced due to network unbalances is a *permanent overvoltage*, with a longer duration and a lower amplitude.

In order to protect our installation against overvoltage, electrical dischargers can be connected at the input and output of each device to be protected.

There are three different protection levels (Fig. 8.9):

Fig. 8.9 Protection level

DEVICE	PROTECTION LEVEL
INVERTER	●
METER	●
CC CABINET	●

- High
- Middle
- Low

Questions

1. Which are the typical electrical values of a PV system?
2. How should be the cooperation between the PV system and the inverter?
3. How must be done the grounding of a PV installation?
4. What kind of issues should include the wiring connections?
5. How must be the building protection of PV modules against high voltage and lightning strikes?

Chapter 9
Connection of the Network

The choice of connecting a domestic photovoltaic (PV) system to a low voltage network should be done in a way to avoid violating the disturbance limits set by the network administrators. In order to avoid shadowing, MV cable will be buried underground, usual voltage will be between 15–30 kV (although it can be a different one depending on each country) and an underground to aerial link will be done, to connect with the power line of the electric company.

The MV cable requires a reinforcement to guarantee that the electrical distribution is homogeneous. This reinforcement is done in three layers (triple extrusion):

- Conductor reinforcement
- Insulation
- Insulation reinforcement

The cable requires also an external cover to provide resistance to humidity; fire; UV sunlight; impact; and chemicals agents (Table 9.1; Fig. 9.1).

9.1 Configuration of the Connection of Building PV Systems—Electrical Network

The protection of the PV module generators in a case of network disruptions as well as the isolation in a case of a complete shut down should be achieved through the automatic switch of the generator or through other suitable protection methods incorporated in the controlling system of the converter so as to avoid damages of the installed equipment and avoid the appearance of risky situations for other users of the network.

E. V. M. Papadopoulou, *Energy Management in Buildings Using Photovoltaics*,
Green Energy and Technology, DOI: 10.1007/978-1-4471-2383-5_9,
© Springer-Verlag London Limited 2012

Fig. 9.1 Underground
wiring

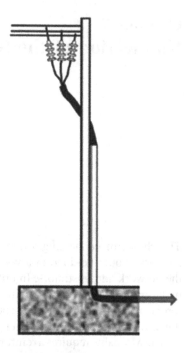

Table 9.1 PV cables

Main features for the copper cable		
Density	g/cm^3	8,89
Resistivity	ohm-mm^2/km	17.241
Conductivity	(% IACS)	100.0
Breaking strength	Mpa	220
Elongation	%	25–30
Corrosion resistance		Excellent

For the secure and proper execution of works through the network, the personnel of electricity suppliers should be provided with the ability of a manual disconnection of the installation from the network, through the free access to the measuring arrangement.

On the other hand, the connection and protection means, should have the ability to disrupt short circuit current voltages and to ensure the direct disconnection of the power generation plant. The regulation of time delay values concerning the protection devices requires special attention because extremely small values may cause an increased number of undesirable disconnections of the production facilities, while great time delays can cause damages to the installed facilities as well to adjacent voltages or producers.

The requirements that must be met for the interconnection of a PV system on the network, in accordance with the instructions of the Network Administrator, summarized as follows (Table 9.2):

Table 9.2 Harmonic distortion and acceptable limits of harmonic constituents' infusion

Parameter	Requirement
Voltage	The value of the AC current to the edges of the electronic inverter should not exceed the −20% (184 V or the +15% of the nominal value of the network voltage. In a case of exceeding the above mentioned limits the disconnection should be made within 0.5 s.
Frequency	The frequency of the output electrical values of the inverter should not be more than ± 0.5 Hz of the nominal value of the network frequency. In a case of exceeding the above mentioned limits the disconnection should be made within 0.5 s.
Automatic reconnection	The reconnection should be made within 0.5 s
Harmonics	The total distortion of the output current should not exceed the 5%.
Infusion of DC current	The maximum value of the infused DC current should be at the most equal to the 0.5% of the nominal values of the facilities.

The high frequency intermittent function of the inverters used in building PV modules causes the appearance of higher harmonic constituents to the waveform of the current that is provided to the electrical network. These higher harmonic constituents are possible to cause problems to the network as well as to facilities that are connected to it and adjacent electronic devices. To be more specific, the infusion of harmonics from the production facilities causes voltage distortion, which directly results in the malfunction of electrical systems (e.g. converters, electrical machines), electronic devices (e.g. network protection systems), as well as adjacent electrical loads (e.g. amplifier, power pack of electronic machinery), which are connected to the same electrical line. On the other hand, the existence of harmonics in frequencies higher than 1 kHz obstructs the use of the network to transfer high frequency telecommunicative signals which serve the duplex data transfer between dispersed energy sources and the control center of the Electrical System. Finally, the existence of higher harmonics may cause the harassment of nearby devices, which are not directly connected to the electrical network (through radiation). This results in the appearance of noise and malfunctions in all the devices, if there is no proper magnetic armor.

To avoid all the above mentioned undesirable situations the compliance of the used converter's function with the existing regulations is necessary.

In detail, the infusion of harmonics from the electronic converters of the building PV modules must be in accordance with the conditions that are predicted by the IEC 61000-3-2 standard. This standard is about the permissible emission limits of harmonic devices and facilities with nominal current lower or equal of 16A/phase which are connected to the X.T networks. Therefore the limits that are set according to this are also suitable for the evaluation of electrical inverters that are used in house PV systems. It must be noted that the control of the harmonic is made only for the standard function of the facilities and not during transitive periods, which usually lasts few seconds (e.g. during parallelism with the network). Finally, the EN 50081-1 put a limit to the permissible radiation emission limits and conductive emissions of electronic converters and the EN50082-1

Table 9.3 Detection of situations of isolated operation—"island phenomenon"

Maximum permitted intensity of odd class harmonics (A)		Maximum permitted intensity of even class harmonics (A)
Harmonic class		The even class harmonics must be at least 25% lower of the corresponding of odd class
3rd–9th	4% of the output current	
11–15th	2% of the output current	
17th–21st	1,5% of the output current	
23–33rd	0,6% of the output current	
Maximum value of T.H.D		5%
Coefficient power on the 50% of the nominal power		0.9
Maximum value of the infused DC current to the electrical network		Less than 0.5% of the facility's nominal output current
Maximum allowed variations of the facility's output voltage in the standard function		85–110% (196–253 V)
Maximum allowed variation of the facility's electrical output sizes in the standard function		(50 ± 0.5) Hz

defines the protection of the mentioned converters from radiation emissions in domestic, commercial, and light industrial environment.

Beyond the above specifications, the Network's Administrator imposes, as necessary requirement for the connection of production facilities to the distribution networks, the achievement of a Total Harmonic Distortion, T.H.D coefficient of the current output lower or equal to 5%, Power Factor, PF higher than 0.95 for inductive and capacity behavior for power over 50% of the nominal and maximum value of the infused DC current (as long as the electronic converters do not have low frequency inverter) at the most equal to 0.5% of the nominal facility current. The goal of these two specifications is the economic operation of the electrical system (through limiting the losses in the network's conductors) and the avoidance of repletion to the network inverters.

In countries with greater experience on the field of PV systems, the limits of the above mentioned specifications are stricter in some cases, due to the increased insinuation of these systems to their energy supply. Next table shows the predicted limits according to IEC 61727 standard which is based mostly on regulations that were formed by the extensive application of PV systems in Germany (Table 9.3).

With the term "Island phenomenon" a non desirable situation in which a part of the electric network which contains both electrical loads and dispersed production units, remains electrified because of the above units while the remaining electric network that is inactive is defined. Causes of this phenomenon may be the deliberate disconnection of a part of the network from the protection means, due to the detection of an error, or a scheduled interruption due to maintenance, or an electricity supply interruption due to external environmental causes, or due to a possible failure of a part of the equipment of the electrical installation system as well as due to human error.

The reasons that impose the detection of such situations are the assurance of a high quality produced power for the consumers and especially the safety for facilities and humans.

In detail, in cases of a scheduled maintenance, while the administrators of the network interrupt the operation of parts of the electrical system in order to make maintenance works, the potential power supply of this section from scattered sources (Due to inability to detect the interruption) puts at risk the personnel performing the necessary work. Additionally, if the protections of an electrical installation system turn on the security switches of a line (Due to detection of random errors, possible equipment damage, external environmental causes, human error, etc.) and it is not practically possible the scattered sources to detect the interruption of electricity supply, they will continue to supply the loads that are connected to the same line with them.

This may cause two important problems:

(a) During the period of interruption, in the part of the deactivated line there is no central control of the frequency and the voltage, and this fact may cause serious damages to other connected users when the scattered sources are not able to supply the loads with the necessary amounts of active and reactive power.

(b) In the case where the scattered production units are able to meet the demands of the loads, when the switches of the protection system reconnect the line with the main electrical network, there may be significant differences between the voltage in the edge of the scattered sources and the voltage of the remaining electrical installation system. (Loss of synchronization with the main electrical network). These differences may have disastrous consequences for both the installed system itself as well as for the other connected consumers that use it.

The inverters of the building PV systems should have protection through the "island phenomenon" with VDE 0126-1-1 or equivalent method or with IEC 62116. In the case of detecting an isolated operation (regardless of the method in use), the disconnection of the PV modules form the electrical network should take place in a time period shorter by 1 s (Time required for the clearance of random not serious errors) so as to minimize the consequences that may result from any rapid restoration of the network voltage.

In Germany, the PV generator is disconnected from the network if the limits for voltage and frequency are violated. The protection from the "island phenomenon" with VDE 0126-1-1 is mandatory for PV systems of power up to 30 kVA only when the connection point of the source with the network is accessible from the network administrator. The compliance with the standard VDE 0126-1-1 is proved by a certificate released by an independent laboratory.

9.2 Safety Instructions

Unlike most power generation plants where the electricity production can be interrupted with the help of a general means of disconnection, the PV modules directly produce voltage in their edges when they are exposed to sunlight. Therefore, installing a PV system usually takes place under voltage conditions from the side of the modules. Also, taking into account that the maximum value of the current of a PV module is slightly larger than the nominal power of the module, it is assumed that the use of safeties does not guarantee the interruption of the system in the case of an error (short circuit of the module). This has as a result that a short circuit error in the side of the DC current may continue to exist regardless of the use of safeties. Good design and proper selection of wiring materials is necessary for security against electric shock not only of the installer but also for any person coming into contact with the system.

Additionally, the selection of wiring with the proper section width guarantees prevention of a fire due to overheating of cables, in case of short circuit.

The PV modules that are going to be used must fulfill either all the technical specifications of the standard EN-IEC 61215 regulation (PV crystalline silicon) either the requirements of the standard EN-IEC 61646 regulation (PV thin-film technology).

9.3 Measures for the Risk Reduction of an Electric Shock During the Installation of a PV System

During the installation of the PV systems, the installer comes into contact with the edges of the module where there is DC current. Usually this value does not exceed the safety limits of continuous contact based to the standard IEC 364-4-41 regulation. The indicative values of the commercial modules' voltage vary between 17 and 100 V (depending on the technology and the numbers of cells). Despite this fact, the electronic inverters that are used in PV systems usually demand the series connection of two or more modules and as a result the series voltage often exceeds the safety limits. The voltage of the series can be calculated if we multiply the number of the series modules with one's maximum voltage.

Consequently, the installation of the system must be performed by authorized staff and according to the following measures:

• *Proposed method of installation*

Significant part of the cabling must be done before the installation of the PV modules. Indicatively, first we put the general disconnection medium of the DC current side and the connection boxes. Afterwards we connect the positive and negative pole of the whole array with the general disconnection medium without performing the intermediate module connections. Then the series connection of the array's modules is performed, while finally the general disconnection medium is connected to the input of the electronic inverter. The proposed methodology aims at avoiding dangerous voltages during installation.

- *Installation with no sunlight*

To avoid the occurrence of high voltages the system installation can be done either by completely covering the modules or at night whenever possible. The use of special gloves and insulated tools is also recommended.

- *Warning signs*

During the installation of the PV system a special warning must be used to warn about the risk of electric shock.

- *Choice of insulation cables and connection boxes*

The use of cables and connection boxes with double insulation minimizes the risk of electric shock. For this reason the use of materials and modules of class II (Class II construction) is recommended. Since the class of materials and modules may not be apparent, the installer must confirm by contacting the manufacturer.

- *Choice of PV modules with preinstalled connection system*

The PV modules with insulated connectors minimize the possibility of the installer being exposed to dangerous voltages. This option is necessary if the installation is done by unqualified personnel.

- *Avoid grounding of the DC current side during installation*

A system in which neither of the two poles are grounded poses fewer risks (compared to a grounded system) because it minimizes the number of possible routes of electric current. For example, suppose in a system grounded to the negative pole the installer comes into contact with any part of the array—and is also in contact with the ground—a current route through him and the ground is created. In such case, the voltage in which the installer will be exposed to equals the sum of voltages of the series connected modules between the contact point and the negative pole of the array.

9.4 Note

It should be noted that the proposed connection methodology cannot fully eliminate the possibility of electric shock.

Minimum necessary equipment requirements
PV modules

- IEC-EN 61215 η 61646,
- IEC 61730—Class A (with Class II insulation)

  The above certificates must always be provided by accredited
  Laboratories

- Electronic converters

  Assurance that they have protection against anti-islanding according VDE 0126-1-1 or equivalent method (type certificate by an independent laboratory)

- Protections of voltage and frequency limits (hypertension, hypotension, hyper frequency—hypo frequency)
- THD output current less than 5%, a certificate of manufacturer compliance (optional)

  If the electronic converters do not have iron inverter then the maximum value of the infused DC current in the electrical network must be less the 0.5% of the value of the inverter's nominal output current, certificate of manufacturer

9.5 Solar PV System Sizing

1. *Determine power consumption demands*

The first step in designing a solar PV system is to find out the total power and energy consumption of all loads that need to be supplied by the solar PV system as follows:

- Calculate total Watt-hours per day for each appliance used.
  Add the Watt-hours needed for all appliances together to get the total Watt-hours per day which must be delivered to the appliances.
- Calculate total Watt-hours per day needed from the PV modules.
  Multiply the total appliances Watt-hours per day times (the energy lost in the system) to get the total Watt-hours per day which must be provided by the panels.

2. *Size the PV modules*

Different size of PV modules will produce different amount of power. To find out the sizing of PV module, the total peak watt produced needs. The peak watt (Wp) produced depends on size of the PV module and climate of site location. Panel generation factor is different in each site location.

To determine the sizing of PV modules, calculate as follows:

- Calculate the total Watt-peak rating needed for PV modules
  Divide the total Watt-hours per day needed from the PV modules
- Calculate the number of PV panels for the system
  Divide the answer obtained in previous item by the rated output Watt-peak of the PV modules available
  Increase in any fractional part of result to the next highest full number and that will be the number of PV modules required.

Result of the calculation is the minimum number of PV panels. If more PV modules are installed, the system will perform better and battery life will be improved. If fewer PV modules are used, the system may not work at all during cloudy periods and battery life will be shortened.

3. *Inverter sizing*

An inverter is used in the system where AC power output is needed. The input rating of the inverter should never be lower than the total att of appliances. The inverter must have the same nominal voltage as the battery.

For stand-alone systems, the inverter must be large enough to handle the total amount of Watts that will be used at one time. The inverter size should be 25–30% bigger than total Watts of appliances. In case of appliance type is motor or compressor then inverter size should be minimum 3 times the capacity of those appliances and must be added to the inverter capacity to handle surge current during starting.

For grid tie systems or grid connected systems, the input rating of the inverter should be same as PV array rating to allow for safe and efficient operation.

4. *Battery sizing*

The battery type recommended for using in solar PV system is deep cycle battery. Deep cycle battery is specifically designed to be discharged at low energy level and rapidly recharged or cycle charged and discharged day after day for years. The battery should be large enough to store sufficient energy to operate the appliances at night and cloudy days. To find out the size of battery, calculate as follows:

- Calculate total Watt-hours per day used by appliances.
- Divide the total Watt-hours per day used by 0.85 for battery loss.
- Divide the answer obtained in item 4.2 by 0.6 for depth of discharge.
- Divide the answer obtained in item 4.3 by the nominal battery voltage.
- Multiply the answer obtained in item 4.4 with days of autonomy (the number of days that the system to operate when there is no power produced by PV panels) to get the required

Ampere-hour capacity of deep *cycle battery*.

Battery capacity (Ah) = Total Watt-hours per day used by appliances
× Days of autonomy (0.85 × 0.6
× no minal battery voltage)

5. *Solar charge controller sizing*

The solar charge controller is typically rated against Amperage and Voltage capacities. Select the solar charge controller to match the voltage of PV array and batteries and then identify which type of solar charge controller is right for your application. Make sure that solar charge controller has enough capacity to handle the current from PV array.

For the series charge controller type, the sizing of controller depends on the total PV input current which is delivered to the controller and also depends on PV panel configuration (series or parallel configuration).

According to standard practice, the sizing of solar charge controller is to take the short circuit current (Isc) of the PV array, and multiply it by

Solar charge controller rating = Total short circuit current of PV array × 1.3

Remark For MPPT charge controller sizing will be different.

Example A house has the following electrical appliance usage:

- One 20 W fluorescent lamp with electronic ballast used 4 h per day.
- One 90 W pump used for 2 h per day.
- One 120 W refrigerator that run 24 h per day with compressor running 12 h and off 12 h.

The system will be powered by 12 V DC, 180 Wp PV module.

1. *Determine power consumption demands*

Total appliance use = (20 W × 4 h) + (90 W × 2 h) + (120 W × 24 × 0.5 h)
= 3'140 Wh/day

Total PV panels energy needed = 3'140 × 1.3 = 4'082 Wh/day.

2. *Size the PV panel*

2.1. Total Wp of PV panel capacity needed = 4'082/3.4 = 1'201 Wp
2.2. Number of PV panels needed = 1'201/180 = 7 modules

So this system should be powered by at least 7 modules of 180 Wp PV module.

3. *Inverter sizing*

Total Watt of all appliances $= 20 + 90 + 120 = 230$ W
For safety, the inverter should be considered 25–30% bigger size.
The inverter size should be about 300 W or greater.

4. *Battery sizing*

$$\text{Total appliances use} = (20\,\text{W} \times 4\,\text{h}) + (90\,\text{W} \times 2\,\text{h}) + (120\,\text{W} \times 24\,\text{h})$$

Nominal battery voltage $= 12$ V
Days of autonomy $= 2$ days
Battery capacity $= \{[(20\,\text{W} \times 4\,\text{h}) + (90\,\text{W} \times 2\,\text{h}) + (120\,\text{W} \times 24\,\text{h})] \times 2 \}/$
$(0.85 \times 0.6 \times 12) = 1\text{'}013$ Ah

So the battery should be rated 12 V 1'020Ah for 2 day autonomy.

5. *Solar charge controller sizing*

PV module specification
Pm $= 180$ Wp
Vm $= 16.7$ Vdc
Im $= 6.6$ A
Voc $= 20.7$ A
Isc $= 7.5$ A
Solar charge controller rating $= (7\text{strings} \times 7.5\,\text{A}) \times 1.3 = 68,25$ A

So the solar charge controller should be rated 70 A at 12 V or greater.

Questions

1. How must the building PV systems to a low voltage electrical network be connected?
2. What is the "island phenomenon"?
3. Refer some tips about PV system sizing?
4. How should the size of battery be calculated?
5. What differences exists between sizing a grid on and stand- alone PV system inverter?

Chapter 10
Epilogue

Energy management embodies engineering, design, applications, utilization, and to some extent the operation and maintenance of electric power systems to provide the optimal use of electrical energy. The most important step in the energy management process is the identification and analysis of energy conservation opportunities, thus making it a technical and management function, the focus being to monitor, record, analyze, critically examine, alter and control energy flows through systems so that energy is utilized with maximum efficiency.

Energy management has become an important issue in recent times when many utilities around the world find it very difficult to meet energy demands which have led to load shedding and power quality problems. An efficient energy management in residential, commercial and industrial sector can reduce the energy requirements and thus lead to savings in the cost of energy consumed which also has positive impact on environment.

In the previous century, the industrial revolution was powered by coal leading to setting up of large power plants as it was the only reliable source of energy available in abundance.

Over the years, oil replaced coal as it was the cleaner form of fuel leading to increased industrialization. Due to increased usage of coal and oil in the name of economic development, environmental problem has started to put a lid on economic progress. The environmental concerns of fossil fuel power plants are due to sulfur oxides, nitrogen oxides, ozone depletion, acid rain, carbon dioxide, and ash. The environmental concerns of hydroelectric power plants are flooding, quality, silt, oxygen depletion, nitrogen, etc.

The environmental concerns of nuclear power plants are radioactive release, loss of coolant, reactor damage, radioactive waste disposal, etc. The environmental concerns of diesel power plants are noise, heat, vibrations, exhaust gases, etc. Finding and developing energy sources that are clean and sustainable is the challenge in the coming days.

E. V. M. Papadopoulou, *Energy Management in Buildings Using Photovoltaics*,   99
Green Energy and Technology, DOI: 10.1007/978-1-4471-2383-5_10,
© Springer-Verlag London Limited 2012

During the last decade a continuously increasing interest in renewable energy technologies was noted in whole world. This was a combined effect of the favorable legal and financial measures that were implemented or the rich potential of Renewable Energy Sources (RES) that exists in many countries and the rising environmental awareness.

Environmental benefits of solar energy are obvious. But there are also psychological and educational advantages of producing electricity in such an easy and direct manner. For many people who have installed their own PV panels, it is a real pleasure to monitor how much solar energy they are harvesting. It can also be an opportunity for them to learn about the differences between energy and power, DC and AC power, voltage and current.

The psychological element is an important but often overlooked explanation of why individuals and small organizations are ready to invest in PV systems even if more profitable financial investments are available.

The main problem today is that renewable energy technologies are too costly. The reduction of greenhouse gas emissions is the final justification for the development of PV, when the electricity produced is so expensive.

All means of electricity generation, including photovoltaic (PV) systems, create polluting emissions when the entire life-cycle is taken into account. In the case of PV systems, those emissions are concentrated in the manufacturing stage. PV manufacturing is energy intensive, resulting in the emissions that accompany the use of standard grid electricity. The energy balance of a PV system is expressed by the *Energy Pay-Back Time* (*EPBT*), which is the time it takes for the PV system to generate the amount of energy equal to that used in its production.

Renewable energy systems that rely on rare metals such as indium or platinum do not have the potential to take us toward a global zero carbon emission economy. That said, all renewable energy systems make use of materials which are available in a more or less limited supply. This does still not mean that their negative impact is as great as that of fossil fuels. Energy use has in general a far greater environmental impact than material use.

And even the development of those renewable energy systems which make use of very scarce materials is not necessarily a wasted effort. Such systems can be quite useful during a transition period and the technical developments that were realized designing them can be the starting point for further development toward more sustainable alternatives.

In view of the limited resources of fossil fuels, the use of renewable energy sources is becoming increasingly important. Alongside wind turbines, photovoltaic systems are a key area of interest. Both the ecological and economic aspects of these systems are of great importance.

PV systems play an important role in CO_2 reduction and also make good business sense, not least in view of the feed-in tariffs guaranteed by local laws. The European Photovoltaic Industry Association (EIPA) forecasts growth of up to 32% in the PV sector by 2013 which has been driven by new governmental promotional schemes introduced in many countries. The construction and operation of photovoltaic systems is now integrated in the system of standards under

IEC 60364-7-712. Compliance with this standard guarantees the safe construction and operation of the PV system.

The advantages of using PV system in buildings are as follow:

- As architectural or promotion element
- Enhances the building aesthetics
- Substitute for other building elements
- Use of synergies—multifunctional building element
- The building envelopes supply sufficient area, thus no additional land used
- Allows to create environmentally benign and energy efficient buildings
- Modular structure of PV—easy application at the place of use
- Energy efficient building: 100% of energy consumption may be covered with PV at 5–10% additional building costs
- Decentralised power source feeds into the building grid
- Solution against black outs, brown outs in combination with a battery bank
- Peak power saving—Improves the building energy management
- High quality power conditions

Generating part or all of a facility's electricity with photovoltaic (PV) systems is growing in popularity all over the world. Whether the systems are used to lower peak demand costs, power an individual facility or enhance the green aspects of a project, even the smallest systems can help lower electric bills and clean up the environment.

Of course, upfront cost remains an issue. Many countries, however, offer a range of incentive programs, many of which could take a PV system from the red to the black in as short a period as 7 years. And as utility rates increase, payback periods shrink.

That does not mean installing a PV system is a simple decision. There is still a lot of planning to do and considerations beyond cost to think about. The process can be further complicated as PVs continue to evolve with improved technology.

Owning a PV system requires a lot of planning and research. And depending on the system, there may even be some maintenance expense. So the question some facility executives are asking—"What do we really want from a green power system?"—We want green power.

Green power rates are competitive and, over the long term, owners save considerably on energy costs, starting in the first year. There is no doubt that the production of electricity with the use of photovoltaic (PV) panels is on the rise. For many companies, large and small, investment in green power is a way to demonstrate environmental commitment and to appeal to green consumers. Green power investments can help improve public relations, differentiate products, and help satisfy corporate social responsibility targets.

With regard to the introduction of Photovoltaic systems; it is hoped that the PV systems on buildings, will have a serious impact on installed PV capacity as the building sector constitutes the "natural environment" for photovoltaic systems.

The photovoltaic technology is suitable for building applications, making it the main component for a decentralized RES development model.

In countries with developed photovoltaic markets, such as Germany, the small PV systems in buildings (<10 kWp) constitute 40% of the annual market, while globally the building sector has a share of the order of 90%.

PV units will become a standard component integrated in the electrical systems of residences and office buildings. PV will also be integrated into the control and information system of smart buildings.

By taking energy from the sun and turning it into useable electricity for a building, we reduce peak energy loads for utilities, and minimize environmental degradation for everyone. Though high initial costs and design constraints have impeded the economic progress of BIPV applications, the economic and environmental attractiveness of Building Integrated Photovoltaics continues to grow. With its multifunctional nature, PV system technology adds a new dimension to the design and construction fields.

Appendix I

A.1 Roof Types

The Gable roof is one of the most popular choices when deciding a style of roof for your home. It has two roof surfaces of the same size, that are pitched at the same angle back to back, making a ridge at the top and forming a triangular roof. Its simple design makes it cheap and easy to build. It effectively sheds water, allows for good ventilation, and typically provides the most ceiling space (Fig. A.1).

The Gable roof is not ideal for high wind areas like the hip roof and is the most likely of roof types to suffer damage, usually with the end wall collapsing due to it not being properly braced (Fig. A.2).

Common variations of Gable roofs:

- *Front gable roof.* The gable end is placed at the front (entrance) of the house. Often used for Cape Cod and Colonial style houses.
- *Side gable roof.* One of the most common roofing styles because of its economy.
- *Cross gabled roof.* Simply two gable roof sections put together at a right angle. The two ridges formed by these gable roofs are typically perpendicular to each other. Lengths, pitches, and heights may or may not defer from each other. Often used for Tudor and Cape Cod style houses.
- *Dutch gable.* A hybrid type of gable and hip roof where a full or partial gable is located at the end of a ridge offering more internal roof space and/or increased aesthetic appeal.

The Mansard roof gets its name from the architect Francois Mansart who popularized it in the 1600s in France. A mansard roof has two distinctly different slopes on each side. The lower portion of the roof has a very steep pitch often with dormers attached, while the upper portion has a low slope, just enough for water runoff to occur. Typically speaking the low slope portion of the roof cannot be seen from the ground.

E. V. M. Papadopoulou, *Energy Management in Buildings Using Photovoltaics*,
Green Energy and Technology, DOI: 10.1007/978-1-4471-2383-5,
© Springer-Verlag London Limited 2012

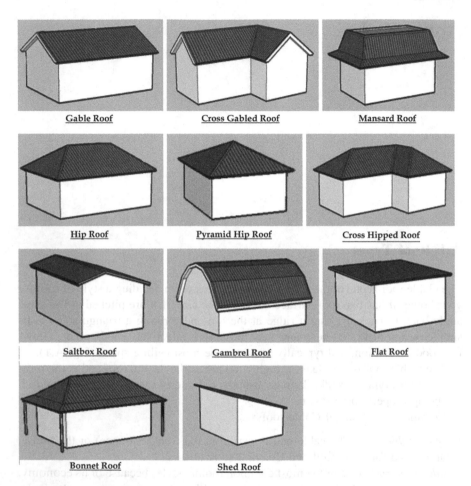

Fig. A.1 Roof types

Mansard roofs offer so much attic space that it is often used as an extra story for the house with the additional space known as "the garret". This type of roof is not recommended for areas that have a great deal of snowfall. Heavy snow buildup could occur on the low slope portion of the roof placing undue strain on the bracing.

Buildings with Mansard roofs (sometimes referred to as Second Empire) enjoyed a popularity in North America in the mid to late 1800s as a part of Victorian style architecture.

A very common roof type, the hip roof (or hipped roof), does not have flat sides like the gable roof; instead all sides of the roof slope down to meet the walls of the house. Building a hip roof is more involving than a gable roof, but building the walls for such a house is actually easier as they are all of the same height.

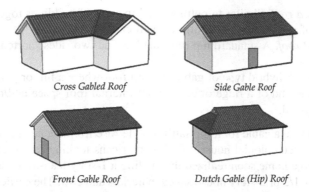

Fig. A.2 Common variations of Gable roofs

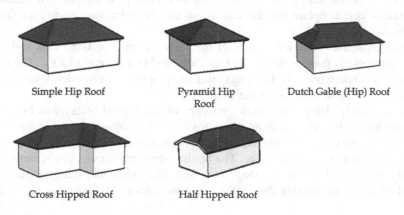

Fig. A.3 Common variations of hip roofs

Hip roofs are very good for homes in high wind or hurricane areas as they offer better internal bracing and are less likely to be peeled from the house as a gable end. Given the roof is at a uniform height gutters can be easily attached around the entire house. Also, the roof protects more of the house from elements such as sun, wind, and rain which over time can require increased maintenance for the structure.

Hip roofs offer less internal roof space making access for maintenance more difficult and offering less potential storage space. Cross hipped (and more complex hip roofs) need to have their valleys kept free from debris so that moisture and dirt do not cause a failure of the valley flashing (Fig. A.3).

- *Simple hip roof.* The most common hip roof with a ridge over a portion of the roof creating two polygon sides and two triangle sides of the roof.
- *Pyramid hip roof.* Four equal triangular sides meet at a single point at the top of the roof.

- *Cross-hipped roof.* Similar to putting two hipped roof buildings together, where the two roof sections meet to form a seam called a valley.
- *Half-hipped roof.* A standard hip roof that has had two sides shortened to create eaves.
- *Dutch gable.* A hybrid type of gable and hip roof where a full or partial gable is located at the end of a ridge offering more internal roof space and/or increased aesthetic appeal.

Most similar to a gable roof, the saltbox style rose from the need to create more space for cramped colonial houses. Early Americans looking for an efficient way to add space to a home soon realized that adding a 1-story lean-to (or shed roof) to the back of a 1 1/2 or 2-story house saved materials and cost. The earliest examples of saltbox houses will sometimes show evidence of the addition by having a second "lean-to chimney/fireplace" or by changing the roof line (slope) on the addition to allow enough height for a useable ceiling. Eventually, the addition became so commonplace that the lean-to was simply added into the original design of the house.

Saltbox houses were a variation of the early Colonial or Cape Cod style and were particularly popular during the late 1600s and into the early 1800s. The name saltbox was taken from the buildings' similarity in shape to wooden-lidded boxes commonly used to hold salt at the time.

John Quincy Adams, the sixth President of the United States was born in a saltbox house that remains standing to this day. (See bottom picture at the right.)

The Gambrel roof, like the mansard roof, has two distinctly different slopes on each of its two symmetrical sides. The bottom slope has a steep pitch, sometimes nearly vertical while the top slope is lower. But unlike the mansard roof, the gambrel roof only utilizes this method on two sides of the structure rather than four.

The gambrel is often referred to as a barn roof as it is commonly seen on many hay barns for the ample space it provides for storage. Small structural additions called dormers may also be seen on gambrel roofs as to provide more head space or extra lighting. This allows for the extra space found in gambrel roofs to be used more effectively.

You can find gambrel roofs on a lot of Dutch Colonial architecture from the 1700s and into the 1800s. The name derives from gamba, a Latin word meaning the leg or hoof of an animal.

Flat roofs usually have at least a slight slope to assist in the shedding of water, thus they are also referred to as "low slope roofs". Flat roofs are typically a more economical roof to build given that it requires less material. While being cheaper to initially build, a flat roof will require re-roofing more often with many materials lasting 10–20 years versus 25–50 years for many pitched roof materials.

Flat roofs are susceptible to failure if pooled water is left for long periods of time. Most recent flat roofs are covered by a continuous membrane to help prevent such water pooling. Yet, flat roofs are not an ideal choice for areas that get a lot of rain and/or snow.

The most common flat roof materials are:

- Roll Roofing
- Built-up roof/Tar and Gravel
- Modified Bitumen
- Rubber Membrane
- Metal Sheets
- Green Roofs

Flat roofs have traditionally been very popular in dry climates including the Southwestern portion of the United States.

Bonnet, one of the least common roofs, could be considered a modified hip roof style. Usually found in French Vernacular architecture bonnet roofs have two slopes on all four sides of a structure. It is essentially the opposite of a mansard roof wherein its upper slope is steeper than the bottom slope. The bottom slope often hangs over the house to cover an open-sided porch and provide shelter from the sun or rain.

Bonnet roofs are sometimes referred to as "kicked eaves" roof. Kicked eaves are considered a "roof enhancement" creating a visor effect to the house.

A shed roof (often called a lean-to) is typically a single roof face that slopes down the entirety of the structure or structure addition. It is a generally the cheapest and easiest roof to build. This roof type is associated with home additions, sheds, and porches. Porches may have open sides, whereas home additions and sheds will usually be fully enclosed.

When used as an addition to a structure, the roof will typically be attached to the building wall on the high side as a lean to. Sheds may also be attached to the house, or may stand alone, and are normally used as a workshop, or for storage.

A shed roof addition coupled with a Cape Cod styled house was used to create the saltbox roof during Colonial America.

Appendix II
Glossary and Acronyms

A

Absorption Coefficient The factor by which photons are absorbed as they travel a unit distance through a material.

Activated Shelf Life The period of time, at a specified temperature, that a charged battery can be stored before its capacity falls to an unusable level.

AIC Amperage interrupts capability. DC fuses should be rated with a sufficient AIC to interrupt the highest possible current.

Alternating Current Electric current in which the direction of flow is reversed at frequent intervals.

AM Air mass; The ratio of the mass of atmosphere in the actual observer-sun path to the mass that would exist if the observer was at sea level, at standard barometric pressure, and the sun was directly overhead. AM0 corresponds to the solar spectrum in outer space, and the reference spectrum for STC was defined to be AM1.5.

Air Mass Approximately equal to the secant of the zenith angle—that angle from directly overhead to a line intersecting the sun. The air mass is an indication of the length of the path solar radiation travels through the atmosphere. An air mass of 1.0 means the sun is directly overhead and the radiation travels through one atmosphere (thickness).

Alternating Current (ac) An electric current that reverses direction periodically.

Ambient Temperature The temperature of the surrounding area.

Amorphous Semiconductor A non-crystalline semiconductor material that has no long-range order.

Amorphous Silicon A thin-film PV silicon cell having no crystalline structure. Manufactured by depositing layers of doped silicon on a substrate.

E. V. M. Papadopoulou, *Energy Management in Buildings Using Photovoltaics*,   109
Green Energy and Technology, DOI: 10.1007/978-1-4471-2383-5,
© Springer-Verlag London Limited 2012

Ampere (A) Unit of electric current. The rate of flow of electrons in a conductor equal to one coulomb per second.

Ampere-Hour (Ah) The quantity of electrical energy equal to the flow of current of one ampere for 1 h. The term is used to quantify the energy stored in a battery.

Angle of Incidence The angle that a light ray striking a surface makes with a line perpendicular to the surface.

Anode The positive electrode in an electrochemical cell (battery). Also, the earth ground in a cathodic protection system. Also, the positive terminal of a diode.

Antireflection Coating A thin coating of a material, which reduces the light reflection and increases light transmission, applied to a photovoltaic cell surface.

Array A collection of electrically connected photovoltaic (PV) modules.

Array Current The electrical current produced by a PV array when it is exposed to sunlight.

Array Operating Voltage The voltage produced by a PV array when exposed to sunlight and connected to a load.

A/E Architectural and Engineering.

Availability The quality or condition of a PV system being available to provide power to a load. Usually measured in hours per year. One minus availability equals downtime.

Azimuth Horizontal angle measured clockwise from true north; 180° is true south.

B

Base Load The average amount of electric power that a utility must supply in any period.

BAS Building Automation System.

Battery A device that converts the chemical energy contained in its active materials directly into electrical energy by means of an electrochemical oxidation–reduction (redox) reaction.

Battery Capacity The total number of ampere-hours that can be withdrawn from a fully charged battery.

Battery Cell The smallest unit or section of a battery that can store electrical energy and is capable of furnishing a current to an external load. For lead-acid batteries the voltage of a cell (fully charged) is about 2.2 V dc.

Battery Cycle Life The number of times a battery can be discharged and recharged before failing. Battery manufacturers specify Cycle Life as a function of discharge rate and temperature.

Battery Self-Discharge Loss of energy by a battery that is not under load.

Battery State of Charge (SOC) Percentage of full charge or 100% minus the depth of discharge. See Depth of Discharge.

Battery Terminology

Captive Electrolyte Battery A battery having an immobilized electrolyte (gelled or absorbed in a material).

Deep-Cycle Battery A battery with large plates that can withstand many discharges to a low SOC.

Lead-Acid Battery A general category that includes batteries with plates made of pure lead, lead–antimony, or lead–calcium immersed in an acid electrolyte.

Liquid Electrolyte Battery A battery containing a liquid solution of acid and water. Distilled water may be added to these batteries to replenish the electrolyte as necessary. Also called a flooded battery because the plates are covered with the electrolyte.

Nickel Cadmium Battery A battery containing nickel and cadmium plates and an alkaline electrolyte.

Sealed Battery A battery with a captive electrolyte and a resealing vent cap, also called a valve-regulated battery. Electrolyte cannot be added.

Shallow-Cycle Battery A battery with small plates that cannot withstand many discharges to a low SOC.

BIPV Building Integrated Photovoltaic; A term for the design and integration of PV into the building envelope, typically replacing conventional building materials. This integration may be in vertical facades, replacing view glass, spandrel glass, or other facade material; into semitransparent skylight systems; into roofing systems, into shading "eyebrows" over windows; or other building envelope systems.

Blocking Diode A diode used to restrict or block reverse current from flowing backward through a module. Alternatively, diode connected in series to a PV string; it protects its modules from a reverse power flow and, thus, against the risk of thermal destruction of solar cells. A diode used to prevent undesired current flow. In a PV array the diode is used to prevent current flow towards a failed module or from the battery to the PV array during periods of darkness or low current production.

Bypass Diode A diode connected across one or more solar cells in a photovoltaic module such that the diode will conduct if the cell(s) become reverse biased.

Alternatively, diode connected anti-parallel across a part of the solar cells of a PV module. It protects these solar cells from thermal destruction in case of total or partial shading of individual solar cells while other cells are exposed to full light.

British Thermal Unit (Btu) The quantity of heat required to raise the temperature of one pound of water one degree Fahrenheit. 1 kW/m^2 Ê 317 BTU/ft2 h.

C

Cadmium Telluride (CdTe) A polycrystalline thin-film photovoltaic material.

CA Commissioning Agent.

Capacity The total number of ampere-hours that can be withdrawn from a fully charged battery at a specified discharge rate and temperature.

Cathode The negative electrode in an electrochemical cell. Also, the negative terminal of a diode.

Cathodic Protection A method of preventing oxidation (rusting) of exposed metal structures, such as bridges and pipelines, by imposing between the structure and the ground a small electrical voltage that opposes the flow of electrons and that is greater than the voltage present during oxidation.

Charge The process of adding electrical energy to a battery.

Charge Controller A device that controls the charging rate and/or state of charge for batteries.

Charge Controller Terminology

> **Activation Voltage(s)** The voltage(s) at which the controller will take action to protect the batteries.

> **Adjustable Set Point** A feature allowing the user to adjust the voltage levels at which the controller will become active.

> **High Voltage Disconnect** The voltage at which the charge controller will disconnect the array from the batteries to prevent overcharging.

> **High Voltage Disconnect Hysteresis** The voltage difference between the high voltage disconnect setpoint and the voltage at which the full PV array current will be reapplied.

> **Low Voltage Disconnect** The voltage at which the charge controller will disconnect the load from the batteries to prevent over-discharging.

> **Low Voltage Disconnect Hysteresis** The voltage difference between the low voltage disconnect setpoint and the voltage at which the load will be reconnected.

Low Voltage Warning A warning buzzer or light that indicates the low battery voltage setpoint has been reached.

Maximum Power Tracking or Peak Power Tracking Operating the array at the peak power point of the array's I–V curve where maximum power is obtained.

Multi-stage Controller Unit that allows different charging currents as the battery nears full SOC.

Reverse Current Protection Any method of preventing unwanted current flow from the battery to the PV array (usually at night). See Blocking Diode.

Series Controller A controller that interrupts the charging current by open-circuiting the PV array. The control element is in series with the PV array and battery.

Shunt Controller A controller that redirects or shunts the charging current away from the battery. The controller requires a large heat sink to dissipate the current from the short-circuited PV array. Most shunt controllers are for smaller systems producing 30 amperes or less.

Single-Stage Controller A unit that redirects all charging current as the battery nears full SOC.

Tare Loss Loss caused by the controller. One minus tare loss, expressed as a percentage, is equal to the controller efficiency.

Temperature Compensation A circuit that adjusts the charge controller activation points depending on battery temperature. This feature is recommended if the battery temperature is expected to vary more than $\pm$ 5°C from ambient temperature. The temperature coefficient for lead acid batteries is typically -3 to -5 mV/°C per cell.

Charge Factor A number representing the time in hours during which a battery can be charged at a constant current without damage to the battery. Usually expressed in relation to the total battery capacity, i.e., C/5 indicates a charge factor of 5 h. Related to Charge Rate.

Charge Rate The current used to recharge a battery. Normally expressed as a percentage of total battery capacity. For instance, C/5 indicates a charging current equal to one-fifth of the battery's capacity.

CHWP Chilled Water Pump.

CHWRT Chilled Water Return Temperature.

CHWST Chilled Water Supply Temperature.

CMMS Computerized Maintenance Management System.

COV Change of Value.

CPU Central Processing Unit.

Cx Commissioning.

Chemical Vapour Deposition A method of depositing thin semiconductor films. With this method, a substrate is exposed to one or more vaporized compounds, one or more of which contain desirable constituents. A chemical reaction is initiated, at or near the substrate surface, to produce the desired material that will condense on the substrate.

CIS Copper–Indium–Diselelide.

Cloud Enhancement The increase in solar intensity caused by reflected irradiance from nearby clouds.

Concentrator A photovoltaic module that uses optical elements to increase the amount of sunlight incident on a PV cell.

Conversion Efficiency The ratio of the electrical energy produced by a photovoltaic cell to the solar energy impinging on the cell.

Converter A unit that converts a dc voltage to another dc voltage.

Crystalline Silicon A type of PV cell made from a single crystal or polycrystalline slice of silicon.

Current (Amperes, Amps, A) The flow of electric charge in a conductor between two points having a difference in potential (voltage).

Cutoff Voltage The voltage levels (activation) at which the charge controller disconnects the array from the battery or the load from the battery.

Cycle The discharge and subsequent charge of a battery.

Czochralski Process A method of growing large size, high quality semiconductor crystal by slowly lifting a seed crystal from a molten bath of the material under careful cooling conditions.

D

Days of Storage The number of consecutive days the stand-alone system will meet a defined load without solar energy input. This term is related to system availability.

DAT Discharge Air Temperature.

DB Dry-Bulb Temperature.

DDC Direct Digital Control.

dP Differential Pressure.

dT Differential Temperature.

DX Direct Expansion.

DC to DC Converter Electronic circuit to convert dc voltages (e.g., PV module voltage) into other levels (e.g., load voltage). Can be part of a maximum power point tracker (MPPT).

Deep Cycle Type of battery that can be discharged to a large fraction of capacity many times without damaging the battery.

Design Month The month having the combination of insolation and load that requires the maximum energy from the array.

Depth of Discharge (DOD) The percent of the rated battery capacity that has been withdrawn.

Diffuse Radiation Radiation received from the sun after reflection and scattering by the atmosphere and ground.

Diode Electronic component that allows current flow in one direction only.

Direct Beam Radiation Radiation received by direct solar rays. Measured by a pyrheliometer with a solar aperture of 5.7° to transcribe the solar disc.

Direct Current Electric current flowing in only one direction.

Discharge The withdrawal of electrical energy from a battery.

Discharge Factor A number equivalent to the time in hours during which a battery is discharged at constant current usually expressed as a percentage of the total battery capacity, i.e., C/5 indicates a discharge factor of 5 h. Related to Discharge Rate.

Discharge Rate The current that is withdrawn from a battery over time. Expressed as a percentage of battery capacity. For instance, a C/5 discharge rate indicates a current equal to one-fifth of the rated capacity of the battery.

Disconnect Switch gear used to connect or disconnect components in a PV system.

Downtime Time when the PV system cannot provide power for the load. Usually expressed in hours per year or that percentage.

Dry Cell A cell (battery) with a captive electrolyte. A primary battery that cannot be recharged.

Duty Cycle The ratio of active time to total time. Used to describe the operating regime of appliances or loads in PV systems.

Duty Rating The amount of time an inverter (power conditioning unit) can produce at full rated power.

DVM digital Voltmeter.

E

Efficiency The ratio of output power (or energy) to input power (or energy). Expressed in percent.

EFG Edge definde Film Growth, A method for making sheets of polycrystalline silicon in which molten silicon is drawn upward by capillary action through a mold.

EIB European Installation Bus which in the meantime is beingused all over the world.

Electrolyte The medium that provides the ion transport mechanism between the positive and negative electrodes of a battery.

Electric Current A flow of electrons; electricity.

Electrical Grid An integrated system of electricity distribution, usually covering a large area.

Electron Volt An energy unit equal to the energy an electron acquires when it passes through a potential difference of one volt; it is equal to 1.602×10^{-19} V.

Energy Density The ratio of the energy available from a battery to its volume (Wh/m^3) or weight (Wh/kg).

EMCS Energy Management Control System

EMS Energy Management System

Equalisation Charge The process of mixing the electrolyte in batteries by periodically overcharging the batteries for a short time.

EVA Ethylene–Vinile–Acetate Foil, it will be used by module production for covering the cells.

F

Fill Factor (FF) For an I–V curve, the ratio of the maximum power to the product of the open-circuit voltage and the short-circuit current. Fill factor is a measure of the "squareness" of the I–V curve.

Fixed Tild Array A PV array set in at a fixed angle with respect to horizontal.

Flat-Plate Array A PV array that consists of non-concentrating PV modules.

Float Charge A charge current to a battery that is equal to or slightly greater than the self discharge rate.

Frequency The number of repetitions per unit time of a complete waveform, expressed in Hertz (Hz).

G

Gassing Gas by-products, primarily hydrogen, produced when charging a battery. Also, termed out-gassing.

Gallium Arsenide (GaAs) A crystalline, high-efficiency semiconductor/photovoltaic material.

GPM Gallons per Minute

Gel Type Battery Lead-acid battery in which the electrolyte is composed of a silica gel matrix.

Grid Term used to describe an electrical utility distribution network.

Grid Connected PV System A PV system in which the PV array acts like a central generating plant, supplying power to the grid.

H

Hybrid System A PV system that includes other sources of electricity generation, such as wind or diesel generators.

HVAC Heating, Ventilating, and Air-Conditioning

HWST Hot Water Supply Temperature

I

I–U Characteristics The plot of the current versus voltage characteristics of a photovoltaic cell, module, or array. Three important points on the I–V curve are the open-circuit voltage, short-circuit current, and peak power operating point.

Incident Light Light that shines onto the face of a solar cell or module.

Irradiation The solar radiation incident on an area over time. Equivalent to energy and usually expressed in kilowatt-hours per square meter.

Insolation The solar radiation incident on an area over time. Equivalent to energy and usually expressed in kilowatt-hours per square meter.

IAQ Indoor Air Quality

Inverter (Power Conditioning Unit, PCU, or Power Conditioning System, PCS) In a PV system, an inverter converts dc power from the PV array/battery to ac power compatible with the utility and ac loads.

Inverter Terminology

Duty Rating This rating is the amount of time the inverter can supply its rated power. Some inverters can operate at their rated power for only a short time without overheating.

Frequency Most loads in the United States require 60 Hz. High-quality equipment requires precise frequency regulation—variations can cause poor performance of clocks and electronic timers.

Frequency Regulation This indicates the variability in the output frequency. Some loads will switch off or not operate properly if frequency variations exceed 1%.

Harmonic Content The number of frequencies in the output waveform in addition to the primary frequency. (50 or 60 Hz.) Energy in these harmonic frequencies is lost and may cause excessive heating of the load.

Input Voltage This is determined by the total power required by the ac loads and the voltage of any dc loads. Generally, the larger the load, the higher the inverter input voltage. This keeps the current at levels where switches and other components are readily available.

Modified Sine Wave A waveform that has at least three states (i.e., positive, off, and negative). Has less harmonic content than a square wave.

Modularity The use of multiple inverters connected in parallel to service different loads.

Power Factor The cosine of the angle between the current and voltage waveforms produced by the inverter. For resistive loads, the power factor will be 1.0.

Power Conversion Efficiency The ratio of output power to input power of the inverter.

Rated Power Rated power of the inverter. However, some units can not produce rated power continuously. See duty rating.

Root Mean Square (RMS) The square root of the average square of the instantaneous values of an ac output. For a sine wave the RMS value is 0.707 times the peak value. The equivalent value of ac current, I, that will produce the same heating in a conductor with resistance, R, as a dc current of value I.

Sine Wave A waveform corresponding to a single-frequency periodic oscillation that can be mathematically represented as a function of amplitude versus angle in which the value of the curve at any point is equal to the sine of that angle.

Square Wave A waveform that has only two states, (i.e., positive or negative). A square wave contains a large number of harmonics.

Surge Capacity The maximum power, usually 3–5 times the rated power, that can be provided over a short time.

Standby Current This is the amount of current (power) used by the inverter when no load is active (lost power). The efficiency of the inverter is lowest when the load demand is low.

Voltage Regulation This indicates the variability in the output voltage. Some loads will not tolerate voltage variations greater than a few percent.

Voltage Protection Many inverters have sensing circuits that will disconnect the unit from the battery if input voltage limits are exceeded.

Irradiance The solar power incident on a surface. Usually expressed in kilowatts per square meter. Irradiance multiplied by time equals Insolation.

I–V Curve The plot of the current versus voltage characteristics of a photovoltaic cell, module, or array. Three important points on the I–V curve are the open-circuit voltage, short-circuit current, and peak power operating point.

J

Joule (J) Unit of energy equal to 1/3600 kilowatt-hours.

Junction Box A PV generator junction box is an enclosure on the module where PV strings are electrically connected and where protection devices can be located, if necessary.

K

Kilowatt (kW) One thousand watts. A unit of power

Kilowatt Hour (kWh) One thousand watt-hours. A unit of energy. Power multiplied by time equals energy

L

Load The amount of electric power used by any electrical unit or appliance at any given time.

LIT Lead Installing Technician

LP Lead Programmer

Life The period during which a system is capable of operating above a specified performance level.

Life-Cycle Cost The estimated cost of owning and operating a system for the period of its useful life. See Economics section for definition of terms.

Load Circuit The wire, switches, fuses, etc. that connect the load to the power source.

Load Current (A) The current required by the electrical device.

Load Resistance The resistance presented by the load. See Resistance.

Langley (L) Unit of solar irradiance. One gram calorie per square centimeter. $1 \, L = 85.93 \, kWh/m^2$.

Low Voltage Cutoff (LVC) The voltage level at which a controller will disconnect the load from the battery.

M

Maintanace Free Battery A sealed battery to which water cannot be added to maintain electrolyte level.

Maximum Power Point or Peak Power Point That point on an I–V curve that represents the largest area rectangle that can be drawn under the curve. Operating a PV array at that voltage will produce maximum power.

MAT Mixed Air Temperature

Modularity The concept of using identical complete units to produce a large system.

Module The smallest replaceable unit in a PV array. An integral, encapsulated unit containing a number of PV cells.

Module Derate Factor A factor that lowers the module current to account for field operating conditions such as dirt accumulation on the module.

MOS-FET Metal–Oxide–Silicon Field effect transistor; used as semiconductor power switch in charge regulators, inverters etc.

Movistor Metal Oxide Varistor. Used to protect electronic circuits from surge currents such as produced by lightning.

MPP Maximum Power Point; The point on the current–voltage (I–V) curve of a module under illumination, where the product of current and voltage is maximum. For a typical silicon cell, this is at about 0.45 V.

MPPT Maximum Power point Tracker; Means of a power conditioning unit that automatically operates the PV-generator at its MPP under all conditions.

N

NEC An abbreviation for the National Electrical Code which contains guidelines for all types of electrical installations. The 1984 and later editions of the NEC contain Article 690, "Solar Photovoltaic Systems" which should be followed when installing a PV system.

NEMA National Electrical Manufacturers Association. This organization sets standards for some non-electronic products like junction boxes.

NOCT Nominal Operating Cell temperature; The estimated temperature of a PV module when operating under 800 W/m^2 irradiance, 20°C ambient temperature and wind speed of 1 meter per second. NOCT is used to estimate the nominal operating temperature of a module in its working environment.

Nominal Voltage A reference voltage used to describe batteries, modules, or systems.

Nominal Voltage A reference voltage used to describe batteries, modules, or systems (i.e., a 12-volt or 24-volt battery, module, or system).

Normal Operating Cell Temperature (NOCT) The estimated temperature of a PV module when operating under 800 W/m^2 irradiance, 20°C ambient temperature and wind speed of 1 meter per second. NOCT is used to estimate the nominal operating temperature of a module in its working environment.

N-Type Silicon Silicon material that has been doped with a material that has more electrons in its atomic structure than does silicon.

O

Ohm (Ω) The unit of electrical resistance in which an electromotive force of one volt maintains a current of one ampere.

One Axis Tracking A system capable of rotating about one axis.

Open Circuit Voltage The maximum voltage produced by an illuminated photovoltaic cell, module, or array with no load connected. This value will increase as the temperature of the PV material decreases.

O&M Operations and Maintenance

Operating Point The current and voltage that a module or array produces when connected to a load. The operating point is dependent on the load or the batteries connected to the output terminals of the array.

Orientation Placement with respect to the cardinal directions, N, S, E, W; azimuth is the measure of orientation from north.

OSAT Outside Air Temperature

Outgas See Gassing.

Overcharge Forcing current into a fully charged battery. The battery will be damaged if overcharged for a long period.

P

Panel A designation for a number of PV modules assembled in a single mechanical frame.

Parallel Connection Term used to describe the interconnecting of PV modules or batteries in which like terminals are connected together. Increases the current at the same voltage.

Peak Load The maximum load demand on a system.

Peak Power Current Amperes produced by a module or array operating at the voltage of the I–V curve that will produce maximum power from the module. See I–V Curve.

Peak Sun Hours The equivalent number of hours per day when solar irradiance averages 1,000 W/m^2. For example, six peak sun hours means that the energy received during total daylight hours equals the energy that would have been received had the irradiance for 6 h been 1,000 W/m^2.

Peak Watt The amount of power a photovoltaic module will produce at standard test conditions (normally 1,000 W/m^2 and 25° cell temperature).

Photon A particle of light that acts as an individual unit of energy. Its energy depends on wavelength

Photo voltaic Cell The treated semiconductor material that converts solar irradiance to electricity.

Photovoltaic System An installation of PV modules and other components designed to produce power from sunlight and meet the power demand for a designated load.

Plates A metal plate, usually lead or lead compound, immersed in the electrolyte in a battery.

Pocket Plate A plate for a battery in which active materials are held in a perforated metal pocket.

Polycrystalline Silicon A material used to make PV cells which consist of many crystals as contrasted with single crystal silicon.

Power (Watts) A basic unit of electricity equal (in dc circuits) to the product of current and voltage.

Power Conditioning System (PCS) See Inverter.

Power Density The ratio of the rated power available from a battery to its volume (watts per liter) or weight (watts per kilogram).

Power Factor The cosine of the phase angle between the voltage and the current waveforms in an ac circuit. Used as a designator for inverter performance. A power factor of 1 indicates current and voltage are in phase and power is equal to the product of volt-amperes. (no reactive power).

PM Preventive Maintenance

Primary Battery A battery whose initial capacity cannot be restored by charging.

Pulse Width Modulated (PWM) PWM inverters are the most expensive, but produce a high quality of output signal at minimum current harmonics. The output voltage is very close to sinusoidal.

Pyranometer An instrument used for measuring global solar irradiance.

Pyrheliometer An instrument used for measuring direct beam solar irradiance. Uses an aperture of 5.7° to transcribe the solar disc.

R

Rated Module Current The current output of a PV module measured at standard test conditions of 1,000 W/m^2 and 25°C cell temperature.

Reactive Power The sine of the phase angle between the current and voltage waveforms in an AC system.

RH Relative Humidity

RAT Return Air Temperature

Remote Site A site not serviced by an electrical utility grid.

Resistance (R) The property of a conductor which opposes the flow of an electric current resulting in the generation of heat in the conducting material. The measure of the resistance of a given conductor is the electromotive force needed for a unit current flow. The unit of resistance is ohms.

Rated Battery Capacity The term used by battery manufacturers to indicate the maximum amount of energy that can be withdrawn from a battery under specified discharge rate and temperature. See Battery Capacity.

Rated Module Current (A) The current output of a PV module measured at standard test conditions of 1,000 W/m^2 and 25°C cell temperature.

RTU Rooftop Unit

S

Sacrificial Anode A piece of metal buried near a structure that is to be protected from corrosion. The metal of the sacrificial anode is intended to corrode and reduce the corrosion of the protected structure.

SAT Supply Air Temperature

Seasonal Depth of Discharge An adjustment factor used in some system sizing procedures which "allows" the battery to be gradually discharged over a 30–90 day period of poor solar insolation. This factor results in a slightly smaller PV array.

Secondary Battery A battery that can be recharged.

Self-Discharge The loss of useful capacity of a battery due to internal chemical action.

Self Distance The rate at which a battery, without a load, will lose its charge.

Semiconductor A material that has a limited capacity for conducting electricity. The silicon used to make PV cells is a semiconductor.

Series Connection Connecting the positive of one module to the negative of the next module. This connection of PV modules or batteries increases the voltage while the current remains the same.

Series Regulator Type of battery charge regulator where the charging current is controlled by a switch connected in series with the PV module or array.

Shallow Cycle Battery A type of battery that should not be discharged more than 25%.

Shelf Life The period of time that a device can be stored and still retain a specified performance.

Short Circuit Current (Isc) The current produced by an illuminated PV cell, module, or array when its output terminals are shorted.

Silicon (Si) A chemical element, atomic number 14, semimetallic in nature, dark gray, an excellent semiconductor material. A common constituent of sand and quartz (as the oxide). Crystallizes in face-centered cubic lattice like a diamond. The most common semiconductor material used in making photovoltaic devices.

Single-Crystal Silicon Material with a single crystalline formation. Many PV cells are made from single crystal silicon.

Solar Cell See Photovoltaic Cell.

Solar Constant The strength of sunlight; 1353 watts per square meter in space and about 1000 watts per square meter at sea level at the equator at solar noon.

Solar Insolation See Insolation.

Solar Irradiance See Irradiance.

Solar Noon The midpoint of time between sunup and sunset. The point when the sun reaches its highest point in its daily traversal of the sky.

Solar Resource The amount of solar insolation a site receives, usually measured in kWh/m2/day which is equivalent to the number of peak sun hours. See Insolation and Peak Sun Hours.

Specific Gravity The ratio of the weight of the solution to the weight of an equal volume of water at a specified temperature. Used as an indicator of battery state of charge.

Stand-Alone PV System A photovoltaic system that operates independent of the utility grid.

Standard Test Conditions Conditions under which a module is typically tested in a laboratory: Irradiance intensity of 1,000 W/m^2, AM1.5 solar reference spectrum, a cell (module) temperature of 25°C, plus or minus 2°C.

Starved Electrolyte Cell A battery containing little or no free fluid electrolyte.

State of Charge (SOC) The instantaneous capacity of a battery expressed at a percentage of rated capacity.

Stratification A condition that occurs when the acid concentration varies from top to bottom in the battery electrolyte. Periodic, controlled charging at voltages that produce gassing will mix the electrolyte. See Equalization.

String A number of modules or panels interconnected electrically in series to produce the operating voltage required by the load.

Subsystem Any one of several components in a PV system (i.e., array, controller, batteries, inverter, load).

Sulfating The normal result of battery discharge when lead–sulfate forms on the surface and in the pores of the active plate material. Sulfation becomes a problem when large crystals of lead sulfate form on the active material as a result of inadequate charging and batteryneglect or misuse. The large sulfate crystals are difficult to decompose under charge and return sulfates back to the electrolyte. This effectively reducesbattery capacity and life. Large sulfate crystals may be detectable by a hard rough surface on the active plate material or a low specific gravityafter an equalization charge. This is called excessive sulfation or "hard" sulfation.

Sun Path Diagram Graphical representation of the Sun's height and azimuth.

Surge Capacity The ability of an inverter or generator to deliver high currents momentarily required when starting motors.

System Availability The percentage of time (usually expressed in hours per year) when a PV system will be able to fully meet the load demand.

System Operating Voltage The array output voltage under load. The system operating voltage is dependent on the load or batteries connected to the output terminals.

System Storage See Battery Capacity.

T

Temperature Compensation An allowance made in charge controllers set points for battery temperatures. Feature recommended when battery temperatures are expected to exceed ± 5°C from ambient.

TAB Testing, Adjusting, and Balancing

Temperature Factors It is common for three elements in PV system sizing to have distinct temperature corrections. A factor used to decrease battery capacity at cold temperatures. A factor used to decrease PV module voltage at high temperatures. A factor used to decrease the current carrying capability of wire at high temperatures.

TU Terminal Unit

Thin Film PV Module A PV module constructed with sequential layers of thin film semiconductor materials. See Amorphous Silicon.

Tilt Angle The angle of inclination of a solar collector measured from the horizontal.

Total ac Load Demand The sum of the ac loads. This value is important when selecting an inverter.

Tracking Array A PV array that follows the path of the sun. This can mean one-axis, east to west daily tracking, or two-axis tracking where the array follows the sun in azimuth and elevation.

Transformer Converts the generator's low-voltage electricity to higher voltage levels for transmission to the load center, such as a city or factory.

Tray Cable (TC) may be used for interconnecting balance-of-systems (BOS).

Trickle Charge A small charge current intended to maintain a battery in a fully charged condition.

TW/THHN may be used for interconnecting BOS but must be installed in conduit—either buried or above ground. It is resistant to moisture.

Two Axis Tracking A system capable of rotating independently about two axes (e.g., vertical and horizontal).

U

Underground Feeder (UF) may be used for array wiring if sunlight resistant coating is specified; can be used for interconnecting BOS components but not recommended for us e within battery enclosures.

Underground Service Entrance (USE) may be used within battery enclosures and for interconnecting BOS.

Uninterruptible Power Supply (UPS) The designation of a power supply providing continuous uninterruptible service. The UPS will contain batteries.

Utility Interactive Inverter An inverter that can function only when tied to the utility grid, and uses the prevailing line-voltage frequency on the utility line as a control parameter to ensure that the PV system's output is fully synchronized with the utility power.

V

Varistor A voltage-dependent variable resistor. Normally used to protect sensitive equipment from power spikes or lightning strikes by shunting the energy to ground.

VAV Variable Air Volume

VFD Variable Frequency Drive

Vented Cell A battery designed with a vent mechanism to expel gases generated during charging.

Volt The unit of electromotive force that will force a current of one ampere through a resistance of one ohm.

Voltage at Maximum Power The voltage at which maximum power is available from a module.

W

Wafer A thin sheet of semiconductor material made by mechanically sawing it from a single-crystal or multicrystal ingot or casting.

Water Pumping Terminology

   Centrifugal Pump See rotating pump

   Displacement or Volumetric Pump A type of water pump that utilizes a piston, cylinder and stop valves to move packets of water.

Dynamic Head The vertical distance from the center of the pump to the point of free discharge of the water. Pipe friction is included. See Friction Head.

Friction Head The energy that must be overcome by the pump to offset the friction losses of the water moving through a pipe.

Rotating Pump A water pump using a rotating element or screw to move water. The faster the rotation, the greater the flow

Static Head The vertical distance from the water level to the point of free discharge of the water. It is measured when the pump is not operating.

Storage This term has dual meaning for water pumping systems. Storage can be achieved by pumping water to a storage tank, or storing energy in a battery subsystem.

Suction Head The vertical distance from the surface of the water source to the center of the pump (when the pump is located above the water level).

Wet Shelf Life The period of time that a charged battery, when filled with electrolyte, can remain unused before dropping below a specified level of performance.

Wire Types See Article 300 of National Electric Code for more information

Watt The unit of electrical power. The power developed when a current of one ampere flows through a potential difference of one volt.

Watt Hour (Wh) A unit of energy equal to one watt of power connected for 1 h.

Waveform The characteristic shape of an AC current or voltage output.

WB WetBulb Temperature

Z

Zenith Angle The angle between directly overhead and the line intersecting the sun. (90° zenith) is the elevation angle of the sun above the horizon.

ZEB Zero Energy Building

ZEH Zero Energy House

Bibliography

1. Lee T, Oppenheim D, Williamson T (1995) Australian solar radiation data handbook, ERDC 249
2. Dones R, Frischknecht R (1998) Life cycle assessment of photovoltaic systems: results of Swiss studies on energy chains. Progr Photovolt Res Appl 6:117–125
3. Kato K, Murata A, Sakuta K (1998) Energy payback time and life cycle CO_2 emission of residential PV power system with silicon PV module. Progr Photovolt Res Appl 6:105–115
4. Frankl P, Masini A, Gamberale M, Toccaceli D (1998) Simplified life cycle analysis of PV systems in buildings: present situation and future trends. Progr Photovolt Res Appl 6:137–146
5. Keoleian GA, Lewis G (1997) Application of life cycle energy analysis to photovoltaic module design. Progr Photovolt Res Appl 5:287–300
6. The Australian Gas Association (2000) Assessment of Greenhouse Gas Emissions from Natural Gas, AGA research paper No. 12
7. Paper 'emissions from photovoltaic life cycles' by environmental science & technology (2008) ACS Publications, London
8. Leonardo energy blog article 'PV Systems: the energy to produce them versus the energy they produce'
9. Article 'Greenhouse gas emissions from energy systems: comparison and overview', Paul Scherrer Institute, 2003
10. Papadopoulou EVM (2010) Photovoltaic industrial systems: an environmental approach/ green energy and technology. Springer Verlag, Heidelberg
11. Markvart T, Castaner L (2003) Practical handbook of photovoltaics: fundamentals and applications. Elsevier, Southampton
12. Lienhard HJ IV, Lienhard HJV (2008) A heat transfer textbook, 3rd edn. Phlogiston Press, Cambridge
13. Falk A (2009) Photovoltaics for professionals: solar electric systems marketing, design and installation. Earthscan Ltd, UK
14. Department of Electrical, Electronic and Control Engineering, E.T.S UNED (2008) "VII Curso de Experto Profesional en Energía Fotovoltaica", UNED, Isofotón y Progensa
15. Patel MR (1999) Wind and solar power systems. CRC Press, Boca Raton
16. Wikipedia, the free encyclopedia, http://en.wikipedia.org
17. SMA product guide 2008/2009
18. SMA measurement accuracy: energy values and efficiency for PV inverters [Technical information]

19. IEE Transactions on Industry Applications (2005) 41:(5)
20. Solar energy forum, http://www.solarweb.net
21. Bonn (2008) Photovoltaics in Greece. EuPD Research, December
22. *Guía de la energía solar*, Dirección General de Industria, Energía y Minas de la Comunidad de Madrid, 2006
23. Non nuclear energy research in Europe—a comparative study. European Commision, Volume 1, EUR 21614/1
24. IEA World energy investment outlook (2003) ISBN 92-64-01906-5
25. European photovoltaic industry association (EPIA), http://www.epia.org
26. PV Status Report (2008) ISBN 978-92-79-07446-2
27. Wenham SR, Green MA, Watt ME, Corkish R (2007) Applied photovoltaics. Earthscan Ltd, UK
28. Luque A, Hegens S, Wiley, Sons J (2003) Handbook of photovoltaic science and engineering. Wiley, USA
29. Akashi Y, Neches J (2004) Detectability and acceptability of illuminance reduction for load shedding. J Illum Eng Soc 33:3–13